MY HUMAN HEART

WHERE SCIENCE AND FAITH COLLIDE

BY RON TESORIERO

Published by Ron Tesoriero

PO Box 6068
Kincumber
New South Wales
Australia 2251

Websites: www.myhumanheart.com
www.resontobelieve.com.au

Copyright© 2021 Ron Tesoriero

The moral right of Ron Tesoriero has been asserted

All rights reserved. No part of this publication, including photographic and digital images, may be reproduced, stored in any retrieval system, or transmitted in any form or by any means, electronic, mechanical or otherwise used without prior written permission of the copyright holder.

ISBN 978-0-6451863-4-5

CHAPTERS

1	I Pray 1		25	From The Mother..................155
2	Another Man Prays 7		26	Contamination In Channel 7 159
3	Can It Be True?......................21		27	The Italian Job 167
4	The Christ Of Limpia.............25		28	Answers In Aramaic 175
5	Statue 29		29	Learning New Things 183
6	Tests 39		30	What Does It Mean?.............187
7	Messages...............................47		31	Genealogy Of Jesus 191
8	Signs 57		32	The Way I See It 201
9	Gifts 61		33	Shroud...................................207
10	Unless I See 65		34	Sudarium...............................215
11	29 Million See.......................69		35	Meeting In Murcia................221
12	Unless I Put My Finger..........77		36	Letter.....................................227
13	Buenos Aires Case 85		37	Coincidence 231
14	Case 19990441/1/2 95		38	I Have Done My Work.........237
15	No Doubt About It 99		39	Mike......................................245
16	Lanciano 109		40	The Final Interview..............253
17	Then It Happened Again.......115		41	Rebuild My Church 259
18	She Saw What She Wanted To See 123		42	Haven't I Heard That Somewhere Before?...............263
19	The Miracle Of Mexico 127		43	Earthquakes..........................269
20	Legnica 131		44	What Does He Want?...........275
21	Mankind's Greatest Intellectual Achievement? 135		45	Gabrielle 279
22	The Origin Of Life...............139		46	Heaven 287
23	The Mega-Million Dollar Question.....................145		47	Sacred Heart.........................295
24	Ex Nihilo..............................151			

CHAPTER 1

I PRAY

'Well, she was just seventeen
You know what I mean
And the way she looked
Was way beyond compare
So how could I dance with another
Ooh, when I saw her standing there?
Well, she looked at me
And I, I could see
That before too long
I'd fall in love with her
She wouldn't dance with another
Ooh, when I saw her standing there
Well, my heart went "boom"
When I crossed that room'

From *'She was just standing there'*

The Beatles

The day was notable from the moment I woke up, but by midnight, it had become one of the most significant of my life. It was the eleventh of May. It was my birthday. It was my brother's birthday. And, it was my parents' wedding anniversary.

It was a Saturday night in 1968 and Sydney seemed the perfect place to be young in. The end of my law degree was in sight. My future lay gleaming before me. I was full of bloom and optimism and confidence. I looked the part too, in my white moleskin trousers, tweed coat and cravat, smoking my pipe whilst driving along in my highly polished VW Beetle in my highly polished shoes. I was on my way to a party. I stopped at a red light and while waiting I caught myself thinking that I had everything I could ever wish for. This thought buoyed me for a few blocks until I realized there was one thing missing and that one thing made me consciously do something I hardly ever did: pray. I prayed a prayer which surprised me then even as it surprises me now.

"Jesus, please help me to find the girl I should marry. She has to be the right one because I only have one shot at it. It would be kind of nice to meet her tonight. It is my birthday, you know. Of course you know. Anyway, it would be a great birthday present. Amen."

About thirty people my age, mostly students, milled around in various parts of the house. A pharmacy student earnestly sifting through the vinyl record collection over by the record player captured my attention. I hadn't seen her before. She wasn't part of the usual crowd. Her dress was unfashionably modest with sleeves to her wrists. She was petite with long hair, a pale central European complexion and an aristocratic posture. I sauntered over as casually as I could and started talking about music, and what she liked. The Beatles, of course. She seemed happy to be in my company which is why I was a little taken aback when she turned and moved off to dance with someone else. Something about her fascinated me and I urged myself to do something. Now. Before it was too late. So I strode over and interrupted her on the dance floor and directed her to the side of the room.
"May I have your phone number?"
"I don't have one."
I took that as a no, and a no, I'm not interested.
Then she said, "But I live somewhere." She gave me her address. "And I have a name. Gabrielle."
The very next day I set out for that somewhere but by the time I knocked at the door of a North Shore house my happy bravado from the previous night had disappeared. A brusque young man, about my age opened the door and glared at me.
"Does Gabrielle live here?" I asked.
"Yes, I'm her brother. I'll get her for you. Gabi!" he yelled, "There's someone here to see you."
I heard matter-of-fact heels come tapping down the dark hallway and Gabrielle emerged into the light smiling, draping a mantilla around

her shoulders. She was surprised to see me and pleased. But also sorry because she couldn't speak to me right then. She was on her way to Mass and was already running late.

"Are you a Catholic?" she asked.

"Uhhhm...Yes." I had been to the finest Catholic schools. I had been baptised and confirmed and raised as a Catholic. My parents were of Italian ancestry. Was that Catholic enough to be Catholic? I probably believed in God. Sort of. But whatever I had learned at home or at school wasn't compelling enough to make me act as if I believed in God. As a university student I had abandoned the practice of my faith save for a few desperate prayers uttered in the face of an impending exam, or for a parking space or, as on the unforgettable evening of my twenty third birthday last night, my prayer for a wife.

"Do you go to Mass?"

"No," I confessed sheepishly, "I haven't been for years."

"Well why don't you come with me?"

So I did. Naturally I did. She had a smile I would have followed anywhere. Even into a draughty stone church to kneel on a fiendishly uncomfortable pew for an hour during which she wouldn't whisper a word to me, never mind look at me. I was forcing myself not to stare at her while she had no difficulty in remaining piously prayerful and completely composed. Something extraordinary settled about her person and made her even more beautiful. She seemed very distant but not unfriendly. It started to feel as if an honour had been bestowed on me just to be able to sit next to her.

In the previous four years I had dated different girls and had never asked or been asked about religion or faith or God. The subject wasn't avoided. It was simply considered unimportant and, to be frank, embarrassing. God didn't manifest as a part of their lives and wasn't part of mine. This experience was very strange because dating and being in a church were not on the same page in my playbook. Marrying in one was though.

For three years I courted Gabrielle. We were engaged and then married on the 26 May 1972 in the chapel at St Joseph's College where I had attended secondary school.

I wasn't ready for marriage. I had just set up my law office in an outlying regional town of Gosford which was a pioneering enterprise in itself. I had $200 in a bank account, a chair, a desk, a typewriter, a telephone and one name, my name, on the door. What I didn't have was a single client. Three hours of every working day were spent driving to and from my home to my office. Eventually, with help from my parents and a loan from the bank, we bought a small house closer to my office and moved north from the big city of Sydney to the Central Coast of New South Wales.

Gabrielle never missed Mass. Not even if living in a new town meant she could no longer walk to church. Not even if I, wedding day long gone, no longer accompanied her. Church and praying were her thing. Work and money were mine. In time more names appeared on my law office door. Tesoriero, Henderson and Cotter soon became a successful legal practice. My expertise was in property law. My feet, and my head, were literally and figuratively on the ground. Gabrielle though always had one foot in heaven.

She had to drive a long way to Mass and although I never went with her I wished it wasn't so inconvenient for her. Urban planners and council development authorities are remiss I thought. They plan for shops and libraries and schools and roads and sports and recreation but fail to consider that in every neighbourhood there is a need for places of public worship. Churches, I had always thought, should be close to the centre of a community, like they were in my ancestors' villages in Italy.

So when the new village of Kincumber was being laid out close to where I lived I took a keen interest in its planning. It was named for the mountain towering over it. By most standards it was hardly a mountain, but it was the highest point in this Central Coast region and overlooked meandering inland waterways and the blue strip of pristine Pacific ocean on the eastern edge of Australia. The plan for Kincumber included all the usual amenities but no allocation for a church. I was disappointed because I wanted to help Gabrielle who was still dutifully traveling such long distances to get to Mass. Since my brothers and I jointly owned some land in the vicinity I suggested we donate part of it for a church. They didn't like the idea. And so it didn't happen.

Years went by. The shopping centre, the library, the community centre were built and at the same time we were building a family. Gabrielle devoutly continued her long expeditions to Mass without me, except of course on the big occasions when our four babies were baptised or someone in the family died. I was a birth, marriage, death and Christmas Catholic.

Then one ordinary day in 1987, so ordinary I don't even remember the date, an uninvited stranger walked into my office and changed my life forever. This time I hadn't prayed. He had though, and whoever had answered my prayer to find a wife all those years ago now answered his.

CHAPTER 2

ANOTHER MAN PRAYS

"When I could not see my way God kept my heart full of trust to make all things come right."

St Mary of the Cross, Mary McKillop, Australia's only saint

I was at the office early. Clients always made appointments. That this one didn't was annoying and curious. A small man is his late fifties shuffled in. He wore a heavy black serge cassock reaching to the floor and a white collar like a bandage holding his head onto the medieval mountain of black fabric below it. One hardly saw men wearing their faith on their sleeves anymore in Australia. He seemed more like a character from a black-and-white Italian film than a Catholic priest, most of whom didn't even wear collars anymore. My days as a Catholic schoolboy hadn't left me with any faith to speak of. They had however left me with indelible vestiges of social decorum. I was indifferent to what these men preached but I was still respectful in a juvenile, if perhaps somewhat patronizing, way.

"Good morning , Father," I said and rose to greet him.

"Ron Tesoriero? How do you do? My name is Father William Aliprandi. I've recently been appointed as the parish priest of Kincumber. My boss," he explained, adjusting his spectacles, "the Bishop of Broken Bay, has instructed me to build a church and a school for the Kincumber parish. But it's a poor one. Mostly working class families, tradesmen; a couple of fellas work in the quarry. A bunch of pensioners too. As for myself…I… er… well I have no money either."

He smiled unapologetically. I leaned back in my chair and awaited the predictable, the inevitable next step in this interchange which, I imagined, one way or another, would be his hand stretched out for a donation.

"Well how can I help you Father?"

"I've found it," he announced in triumph.

"You've found the money?"

"Not the money. I've found the land. It's the perfect place. I walk past it every day and it's the right land. At the moment it doesn't look like much: thick bush and there's only a very basic dirt track on the one side. But once it's cleared it will be ideal because it's near the village. So I prayed for it."

"Is it for sale?"

"No."

"Do you know who owns it?"

"No. No, but I…"

"So, let me get this straight. You don't have any money and you want to buy some land which is not for sale. You don't know who owns it. You don't know its size. You don't know its zoning which means you don't know if you can, by law, build a church on it. But you have asked God to make it happen."

"Yes. That's correct. But there's more," the priest continued unfazed by my tone. "You see, every day I've been praying about this whole business of building a church and I believed really strongly that this was the right place for it and so I walked to the middle of it, made a cross with some stones on the ground and now every day I go there and I kneel next to that cross and I pray to God that we might obtain it."

Nothing in all my years in the legal profession had prepared me for the priest's naivete and the astonishing confidence with which he proclaimed his superstitions. I, quite unusually, had absolutely nothing to say. Father Aliprandi wasn't deterred by my silence. He was on a roll.

"I have some help though."

"Good. Very good. Help is good."

"I've enlisted the help of a woman. I have known her to be very influential."

"Excellent. Who is she?"

"Was," he corrected me with a chuckle. "Who was she? She's not quite with us... er... so to speak. Been dead some seventy years. Maybe more; 1910 she died. I think."

My hands, folded in calculated professional composure jerked apart. I reached for my pen and started fidgeting while my brain raced for explanations. Maybe he's come into an inheritance? Or a trust? But seventy years after the fact? Hmmm. Unlikely. I steeled myself with polite banality, "Dead?" I said.

"Well yes. Although I prefer to think of her as living somewhere else. Being dead is such a horribly final thing. Don't you think?"

It was not a question for a lawyer and I gave no answer. He continued unabashed.

"Yes. Well... er... her name is Mary Mackillop. She was a nun, the founder of an order of teaching nuns in Australia, the Sisters of St Joseph of the Sacred Heart. She's Australia best bet of one day having its very own backyard saint. Exceptional woman. Tough as nails. Always had a soft spot for poor children. She founded the old orphanage down in South Kincumber on the water's edge, next to the little stone church; the second oldest in Australia I'm told;1842. She even slept there. Anyway we're all praying that the Holy Father will soon beatify her. But I'm rambling on a bit aren't I?"

With no encouragement from me to continue he wrapped up his history lesson.

"Well all I've done is nail a relic of hers to a tree on that land, and I'm praying for her intercession, so we can have it. She taught kids their faith. She'll want this."

"A relic?"

" Yes. Oh, it's not much. Just a small piece of one of the habits she used to wear."

The whole conversation made me uncomfortable. This was decidedly not what I was used to. In my world reality was neatly compartmentalised. Praying and faith were on one side, the Sunday side, and land conveyancing, ordinances, development applications and planning resolutions were on the other. In a gesture enforcing my world view I swept the nonsense away with a large map rolled into a baton which I brandished like a sword in front of the little priest.

"Let's have a look then, shall we?" I said in my getting down to business voice. "This is a Gosford Council Zoning Map and you should be able to locate the land in question. So have a look and see if you can identify it."

The priest rose to take in the scale of the diagram which encompassed the whole desk and fell with rolled edged down the side. Without any trouble he pointed out a two centimetre square, coloured orange and outlined in black.

"This is it? You're sure?" I asked

"That's it. Maps make the world seem so small and manageable don't they?" he mumbled.

I scrutinised it and decoded the fine print and relieved to be able to dismiss this whole silly episode declared a final verdict.

"Father I am sorry but you will never be able to build your church on this property. It has been zoned as 'Reserved for Conservation.' You are simply not allowed to build a church or a school on it. In fact no-one can build anything whatsoever on it. Not even the Pope."

The victory for the real world and its august regulations, unshakeable by priests or their archaic superstitions, their prayers or their saints, was not as satisfying as I hoped. Father Aliprandi was silent. His gaze dropped. As I rose to bid him goodbye a sensation of pity propelled me to say, "Let me see what I can do. I'll make some enquiries. See if there's anything suitable, with the proper zoning. Just leave your details with my secretary."

"Thank you kindly. Thank you. May the good Lord bless you for your help."

He left, and with him, so did my thoughts about his land and his prayers.

A few days later another client was in my office. He was a wealthy businessman (with an appointment) who, for tax reasons, quickly had to sell two properties before the end of the financial year. The first was desirable enough to already have an interested buyer, a big property developer from Sydney. But the second was a problem. So much of a problem that he was worried he wouldn't be able to offload it in time.

"What's the problem with it?"

"Problems. Plural. It's not worth much because there's no decent access. No roads or services. No electricity, piped water. All that sort of stuff. So I was hoping to somehow force the buyer of the good property to have to buy it as well. 'Force' is not a good word. Induce. Yes, induce the buyer. I'll sell it for a song."

"Okay then. Let's start at the beginning. Let's see what you're allowed to build on this land. Can you point out the property for me on this zoning map? Out came my big map again.

His finger hovered, then rested on the spot. I slumped back into my chair, stared blankly ahead and sighed deeply enough for him to ask,

"That bad eh?"

"No, no… it's just… er… its…"

"Hold on a second. Could I just check something?" the businessman stood up and bent forward to examine the map. "Hmmm. I thought so."

"What?"

"This is not the most current zoning map. They obviously haven't published an up-to-date one yet, the latest one which shows all the newest zoning regulations that have just gone through Council. This Kincumber land of mine is no longer zoned for conservation. It's now zoned residential. So that's some good news surely? Much less restrictive."

I knew that what he was saying was significant but his words had slid into the background as if we were in a movie in slow motion and the sound had been switched off. I tried really hard to bring myself back into real time but all I could think about was that somewhere on his land was a tree with a scrap of a dead nun's clothing nailed to it.

The series of unusual events in my office continued that Friday. The businessman was there to sign the deal on the sale of the first piece of land, his prime land. The interested buyer, a big time property developer, and his lawyer had driven up from Sydney for what I expected would be a purely procedural meeting. I was looking forward to an early lunch. Then out of the blue my client suddenly pulled the rug out from under all of us and said,

"I've changed my mind. The price is too low. It's worth more. I want more."

"But we agreed. We had a deal." The Sydney buyer was livid.

"I know, "said my client smoothly, "but I haven't signed a thing yet. And I've changed my mind. I want more. Two hundred thousand dollars more."

"What! This is outrageous! I've been in this business for thirty five years and I've never had to deal with anything like this." He was shaking his head in disbelief. One man's obstinacy confronted another man's indignation and the hostility in my office was flammable. He stood up to leave. I was as unprepared for my client's reversal as was the buyer. I vacillated between anger at being professionally compromised and a vague apprehension that there was a lot to lose if this deal went sour. In the heat of their hostility I searched in the remote recesses of my legal memory for anything that could salvage the negotiations.

"The land we're dealing with" I said, " is zoned part residential and part conservation, right? The Council regulations permit a certain planning elasticity regarding the development on the conservation section of the land if that land shares similar characteristics with the adjoining residential section which you want to develop. As long as environmental factors are not infringed and as long as they can see the congruence of vegetation, they allow for some encroachment."

Three angry men glared at me, but I concluded by looking exclusively at the very annoyed Sydney developer. "What that means for you is that your development gets to encroach on conservation land overlooking the ocean. You get some pretty amazing sea views as a bonus."

He looked at his attorney and then at me suspiciously.

"No. It's off. The deal's off." He stormed out of the office followed by his attorney hastily grabbing his documents. If he'd had a free hand he would have slammed the door.

If I had been under any illusions that the day had exhausted its quota of the unexpected I was wrong. We had lost one buyer but maybe there was still hope with a second. I took the businessman out to lunch. There was someone I wanted him to meet. Father Aliprandi was already seated when we made our way to our table. I noted the effect the priest's unusual attire had on the businessman. After introductions I came straight to the point.

"Father Aliprandi wants to build a church and a school on your Kincumber land."

"That suits me perfectly. I'm all set to sell immediately."

"Now, now," said the priest, "slow down a bit. I'm not the man with the cheque book. The Bishop still has to approve the purchase and then we'll have to start raising money. It may be a while yet. Lots and lots of cake sales still to be had."

Twenty minutes ago the businessman had shocked me with an extravagant demand for one piece of land and now he shocked me by veering in the opposite direction completely, all the way to extravagant generosity.

"I will give it to you for $210,000. Ron will bear me out when I say it's an absolute steal at that price. A fraction of the market value."

"Thank you so much. That's unbelievably low. You are a very generous man. But bargain or not, I still don't have the money or the permission right now. My superior will be most interested, I know. But we just need a little time."

"I don't have time. I can't afford more time."

The priest was downcast. The businessman grimaced, adjusted his serviette and then announced. "Look here. I'll sell you an option on the land for a dollar."

"A dollar! You mean, one dollar? Singular? Seriously?"

"That's right. You heard me. A dollar. And when you get the rest of the money we will settle."

Father Aliprandi broke out in hearty laughter. "Well now, let me see…" He fumbled in his pockets and drew out odd coins. "Eighty cents. I owe you twenty."

"Done."

We all laughed.

The two men shook hands on the deal. I raised my glass.

"To the church and the school."

Father Aliprandi turned to my client and said in a quiet solemn tone, "For such a commitment to the things of God, don't be surprised if you're blessed by him and good things happen to you. Even this very day."

An hour or so later my client and I were back in my office redrafting the contract of sale for his premier property to include his additional $200,000 in the asking price. There was a knock on the door. My secretary peered around the door.

"There's someone here I think you want to see." This certainly was no ordinary day. The Sydney boys were back. This time the big time developer was quiet. His lawyer did all the talking about using their time proactively by going over to the council offices and asking the planning executive himself about the obscure clause I had mentioned and then reconsidering the deal in terms of it being verified and the encroachment possibility of additional allotments with sea views and (this was the bolt from the blue), that his client had agreed to the higher asking price even though the already steep price was now preposterously inflated.

When he finally stopped talking there was complete astonished silence.

The businessman just sighed, and signed. It was over. Both pieces of land were sold.

I was back in my office before everyone else early on Monday morning. The phone rang. I let it. I was too busy and it was still before eight. Shrill and demanding it continued. I glared at it. It took no notice. I relented. It was my client the businessman.

"Ron, it's unbelievable. It's absolutely unbelievable!"

"What is?"

"You haven't heard? Don't you listen to the news? The market's crashed. Big time. All over the world."

"The stock market?"

"Of course," he said exultantly. "Which means that I would never be able to get the same price for my land today as I got on Friday. It's like winning the lottery. How lucky is that?"

"Luck?" I murmured thinking of the priest, thinking of the market's collapse, thinking of the millions who would suffer. It was 11 October 1987, a day to which very few would attach the word 'luck'. 'Luck' seemed to be inept in explaining the fascinating sequence of events unravelling before my very curious eyes. I wasn't sure what was happening but I was sure 'luck' didn't quite get the full measure of it.

Months passed. I carried on working hard at my uneventful job. Father Aliprandi carried on working hard to raise $210,000 minus one dollar. He was nowhere close to his target when another unforeseen sale of land took place. This time the land in question lay directly next to the Aliprandi land. And this time the purchaser was the New South Wales Government itself. The State had decided that a new high school was to be built. That meant they would first be providing infrastructure for water, electricity, sewerage, telecommunications and tarred roads. This was another extraordinary piece of 'luck' for the priest. He now not only had his land, he also had free government supplied infrastructure to service it. It was totally unexpected very, very good news.

But I started worrying. If Father Aliprandi didn't come up with the outstanding balance he owed for the land he might lose out. I had seen how unpredictable the businessman could be. He may very well prefer the extra million dollars he would most certainly now get for the land to keeping his promise, which after all was only sealed by a handshake.

I explained to the priest.

"Father, if you can't get the money, all of it, soon, you may lose out completely"

"So be it. I don't have it. I can't borrow it from the bank."

Why I did what I did next I can't quite fully explain. It felt like a gamble but there were no stakes.

"I will lend it to you," I said.

Handshakes don't hold up well in court and we had nothing on paper binding the businessman to the offer made all those months ago in the restaurant. It was a deal made in person and so approaching him in person with a cheque for the balance seemed the best way to proceed. But there was by that stage a new obstacle to negotiate, a thousand mile high obstacle. The businessman had packed up and moved way up north to the Gold Coast.

I didn't have time to fly a thousand kilometres to another city. I was busy with my real clients, the ones who made appointments. Where was I to find the time in my calendar for what was essentially an act of charity?

The following morning I walked in to my office to find an airline ticket on my desk. I was to depart for the Gold Coast the following day to conduct business for another client, one who made appointments, someone then completely unrelated to all of this. I shook my head in disbelief. Here in my very hands was a ticket to the very same city paid for by a client. That client happened to be my neighbour, a well known television journalist, Mike Willesee.

I flew to the Gold Coast, completed Mike's business at the architectural firm and then asked at reception to use the phone. I phoned the businessman. He was surprised to hear from me and even more surprised to learn I was in town. I was surprised he was free at that time and even more surprised that he was willing to meet with me. He gave me his address. I was in a strange city and didn't have a clue where he lived or how far away he was in relation to me.

"Where are you right now, Ron?"

I told him the name of the architect's office.

"It's probably easier if I just meet you there then. Is that alright?"

It suited me perfectly. The foyer was fashionably decorated with uncomfortable armchairs and glass tables. I settled into one with a newspaper. No sooner had I read the headlines then the businessman walked through the door. He lived, by sheer coincidence, or was it luck, no more than a block away.

He was not alone. With him was an elegant woman with a sour expression on her polished face. She was not pleased.

"This whole thing is insane. Absolutely ridiculous," she grumbled.

"Honey, we've talked about this," he said to pacify her.

She turned fiercely to address me but it was clearly not me she was trying to convince.

"The government is about to build beautiful brand new roads with electricity and everything. It's worth twenty or thirty times as much now. We have absolutely no reason to accept this ludicrous amount for the land."

"Let's not go through this again," said her husband.

"We are perfectly entitled. We are completely within the law to sell it for its true value. Is that not so Mr Tesoriero?"

I was floundering, searching for some clever way to make the priest's case, some way to charm the offensive, but I had nothing to say except, feebly,

"We have an agreement."

"Not on paper. Not in the eyes of the law. You have no signatures. And you're not getting mine."

I didn't want to be the umpire in this matrimonial clash. I didn't want to alienate the businessman by offending his wife. I didn't want to disappoint the priest. The businessman looked at me with a penetrating directness. Then, without a word, took the document I had laid out and signed it. Still without a word he gave the pen to his wife and pushed the paper in front of her. The gesture, silent and deliberate, was a final verdict in their private married world. She sighed and she signed. We stood to shake hands.

"I gave my word to that priest," he said. " I am going to honour it."

Whether I was more relieved or surprised is debatable but I was silently celebrating. Finally. It's all over.

As I look back now I shake my head and smile to myself. How many times would I say this exact thing in the years which lay ahead? How many times would this exact refrain, 'it's all over', prove to be completely wrong? Each time I thought it was over it was just the beginning.

I immediately phoned the priest.

"Incredible! Unbelievable! In spite of all the odds and in the most convoluted unorthodox way you have your land for a fraction of its true value, and with great spanking new infrastructure thrown in for nothing. But tell me honestly Father, you have three university degrees. Did you really believe when you knelt down on that land like a peasant, looked up to the sky, spoke to a God you can't see and asked for the impossible, that it would actually happen?"

"I had no doubt."

"How could you be so certain?"

"The God who created the world from nothing has not lost his power you know. I've offered my life to work for him and I asked for his help to do that work. I would have been surprised if he had not helped me."

"God deals in real estate? Seriously? He gets involved in details of the here-and-now stuff of business and law? Could this just be a freak series of interconnected coincidences with a religious flavour? I mean, every day someone gets lucky and someone loses. That's just life. Surely?"

He was quiet for a few deliberate seconds then said, "After living a life in which prayers are answered, mine sometimes, and those of others, it would be profoundly ignorant of me to insist on calling them coincidences, don't you think? And inexcusably ungrateful too. I have prayed not to any old gods, but to the God of Abraham, Isaac, and Jacob, the one Jesus called Father and this God answered, intervened, and showed his hand in my life and the lives of people around me. This God is real. "

"Well," I sighed, "would you call it a miracle?"

"Oh no, Ron," he chuckled, "not a miracle. That's something else entirely."

CHAPTER 3
CAN IT BE TRUE

For days afterwards I thought about all the small details that had fallen into place at just the right time in the Aliprandi land acquisition. The cascade of coincidence was extraordinary and yet I was unconvinced by Father Aliprandi's explanation that God had intervened to answer a prayer. In my very secular business world you were regarded as dimwitted if you believed in God. Be that as it may, the questions pestering me did not subside. Does God exist? Does God answer prayers? Does He intervene?

I didn't have the answers and the questions themselves gave me no rest. I began in earnest to explore what the Catholic Church had to offer regarding God's existence, his intervention in the world, and even the suspension of the natural physical laws of the world.

I read about the bodies of saints which remained incorrupt centuries after their death. I read about apparitions of Jesus and his mother, about springs of water which suddenly gushed forth and healed the sick, about divine dictation, about people who seemed to exist in two places at the same time, about people whose hands and feet bled spontaneously without cause. I read about statues which weep and bleed. I read about illiterate Portuguese children who claimed the Virgin Mary appeared in 1917 to warn them of an impending war.

I had always thought that these were fanciful reports from distant cultures in the distant past, but I was wrong. Some had purportedly occurred in my very own modern twentieth century. And some were supposedly happening right now but were so unusual that no mainstream media ever bothered reporting on them. And when on rare occasions they did, the methodology was so unprofessional as to be infuriating, even to someone as inexpert in these matters as myself.

The headline of one, for instance, caught my eye. It read: 'A bloody miracle? No! Statues can't weep.' A statue of the Virgin Mary was reportedly weeping blood in Civitavecchia, near Rome. The newspaper covered the story by approaching a research chemist at the university to replicate the effect of blood streaming from another statue's eyes by tampering with it. The whole article rested on the assumption that the phenomenon was an elaborate hoax. No elementary questions were asked, questions I would have imagined to be standard operating procedure for any investigative journalist: What actually happened? In what circumstances did the statue 'weep'? Were there any witnesses? Was the event recorded on film? Had the film been forensically assessed? Had the statue itself been examined to see if it had been tampered with or engineered to produce droplets of blood? Was it tested in a lab? Was it actually blood? If so, was it human blood? Is it still happening?

The dilemma for the media seemed to be that the intellectual agenda had already been mandated by a pervasive contemporary ideology which decided that religion is for the ignorant, the weak, the fearful and the gullible, for people clinging to the past. It warranted no more than derision and condescension. The narrative framework had been deliberately constructed to be one of alternatives; faith versus reason, science versus religion, intelligence versus ignorance, today versus yesterday, us versus them. To report a story such as this with anything other than a smirk on your face would be to commit professional journalistic suicide. For so much as merely entertaining the possibility that a statue might actually be exuding blood a journalist would be pilloried. Stories of the supernatural were sanitized for contemporary consumption by association with the past, or patronisingly, with people considered to be gullible and feeble-minded, or even, arrogantly, with remote undeveloped places of the world, places like Cochabamba.

In the middle of South America is Bolivia, and in the middle of Bolivia is Cochabamba. It's a small city at the foot of the mountains. Prior to colonization in 1542 by the Spanish, it was inhabited by the Cechna Indians. A gigantic statue of Christ, taller even than the familiar Rio de Janeiro landmark of Cristo Redentor on Corcevado, towers over the city and over a modest house to which I had travelled thousands of miles for the sole purpose of getting some answers. I had no confidence in the standard biased media reporting and wanted to see for myself exactly what was going on with another statue. It too was a statue of Christ.

This one however was small, a bust only 30cm high and made of plaster. It was on a table surrounded by flower arrangements in an otherwise nondescript room filled with people. The upturned eyes and open mouth depicted intense suffering, as if the subject was frozen in the moment of an agonizing cry. Dark streaks followed the contours of the face and an array of encrustations of dark, rough, red-brown material covered sections of the forehead. It was an unnerving sight; gruesome for being so life-like. Someone said that the statue seems to be etched in sorrow with the paint of its own blood.

And this exactly was why I had travelled so far. I had to see for myself, ask my own questions, do my own investigations around the incredible, illogical, bizarre claims that on 9 March 1995 this little statue began to 'weep' real tears and 'bleed' real blood. I had the chance to do what the media had failed to do.

CHAPTER 4
THE CHRIST OF LIMPIAS

"I believed that it was now my duty to swear on oath to what I had seen."

Dr Heriberto de la Villa

When first bought, the statue was one of hundreds exactly like it sold as Santo Cristo de Limpias, the Christ of Limpias. In fact it was a statue which combined two traditional images in one. Yes, it had the upturned crowned head in agony depicted in the Limpias Christ hanging on a cross. But it also, by virtue of the scarlet cloak around its shoulders, depicted another scene in Christ's life, a scene in which he is called 'The Man of Sorrows', prophesied by Isaiah centuries before, a man abandoned, tortured, dejected, and utterly humiliated by the powers that be. In the gospels, Pontius Pilate regards his pitiful state and exclaims disdainfully, *"Ecce Homo,"* (Behold the Man).

The original Cristo de Limpias statue in Spain, on which the head was modelled, is not a bust. Nor is it small. It is a six foot life-sized corpus on a cross suspended high above the sanctuary in the Church of St Peter in the tiny village of Limpias. It garnered attention when in 1919 over 8000 witnesses testified in writing, 2500 of those under oath, that the statue moved, changed colour and facial expression, appeared to perspire, weep and bleed. Among the many witnesses were doctors and physicians. Their published accounts are fascinating because I, like many people, grant a degree of reverent authority to men of science, trusting that their opinions will be less tainted by subjectivity and imagination and motivated more by empirical observation.

The language they used was typically anatomical, scientific, technical and analytical. Dr Eduardo Perez described the 'body' as a real human male in the final moments of his life before he finally died. He wrote what he saw on 6 October 1919.

"...there ensued a violent inhalation with the straining of the muscles of the neck, whereby the musculus sternocleidomastoideus specially stood out and furthermore the musculii pectoralis, the scalenus anterior, and the accessory respiratory muscles contracted, with a considerable dilation of the intercostal spaces, as in the case, for example, at the last struggle after mortal wounds. For a moment he appeared on the point of death...... I must add that during the whole of that afternoon I saw the figure a reddish colour. The following day it was a yellowish or lead colour, as with a dying person."

In May 1920 another doctor, Dr Penamaria, published his account in La Montana. He wrote that he was a sceptic intent on finding a scientific explanation to disprove what he considered must be "hysteria". After seeing the statue move he changed his location in the church a few times to verify that what he was seeing was not a trick of the light from a certain angle. The experience was unnerving enough to prompt him, self-avowedly cynical of religion, to pray. He prayed for even more distinctive proof, for something more extraordinary, something, "that would leave no scope for further doubt and would give me positive grounds for this miracle so that I might proclaim it to all and sundry and defend it against every opponent, even at the risk of losing my own life."

What occurred moments later impelled him to write:

"...A moment later His mouth was twisted sharply to the left, His glassy, pain-filled eyes gazed up with the sad expression of those eyes that look and yet do not see. His lead coloured lips appeared to tremble; the muscles of the neck and breast were contracted and made breathing forced and laboured. His truly hippocratic facies showed the keenest pangs of death. His arms seemed to be trying to get loose from the cross with convulsive backward and forward movements, and showed clearly the piercing agony that the nails caused in his hands at each movement. Then followed the in-drawing of a breath, then a second ... a third ... I do not know how many... always with painful oppression, then a frightful spasm, as with someone who is suffocating and struggling for air, at which the mouth and nose were opened wide. Now follows an outpouring of blood, fluid, frothing, that runs over the under-lip and which he sucks up with his bluish, quivering tongue, that he slowly and gently passes two or three times in succession over the lower lip; then an instant of slight repose, another slow breath ... now the nose becomes pointed, the lips are drawn together rhythmically and then extend, the bluish cheek-bones project, the chest expands and contracts violently after which his head sinks limply on His breast, so that the back of the head can be seen distinctly. Then he expires... I have tried to describe in outline of what I saw during more than two hours."

In another newspaper, Del Pueblo Astur, an atheist, then a medical student now Dr Heriberto de la Villa, published his account on 8 July 1919. "Auto-suggestion" he declared, "is quite out of the question, for I did not believe in the miracle when I went." He was accompanied by another doctor and three other adults. His companions saw the miraculous movements but he did not, even though he moved from one position in the church to another. Even so he was persuaded to return to the church later the same day. He reluctantly did. This is what he saw:

"...little by little the breast and face became dark blue, the eyes move to the right and left, upwards and down, the mouth opens somewhat, as if He were breathing with difficulty. This I saw for fifteen to twenty minutes.

I also noticed that above the left eyebrow a wound formed, out of which a drop of blood flowed over the eyebrows, and remained stationary by the eyelids. After that I saw another drop of blood fall from the crown of thorns and flow over the face. I could distinctly discern it, for it was very red and contrasted with the dark blue colour of the face. Then I saw a quantity of blood drip from the crown of thorns onto the shoulder, but without touching the face. He opened His mouth wide, out of which a white matter like froth welled.

He opened His mouth again, out of which froth and blood streamed in great quantity and flowed out of the comers of the mouth quite distinctly... Thereafter I believed that it was now my duty to swear upon oath to what I had seen and I did so in the sacristy of the church."

Now, some seventy years later, on a different continent, some of the same inexplicable phenomena were being attributed to a little plaster copy of the Christ of Limpias.

CHAPTER 5

STATUE

'My eyes are full of tears my heart of grief.'

William Shakespeare

My legal mind insisted on the value of evidence. To this end I came prepared with camera equipment and a crash course in cinematography from a professional cameraman under my belt.

I was there. I filmed the statue. I filmed it 'weeping'. And then I interviewed other people, who like myself, had seen this unfathomable phenomenon with their own eyes. These are the transcripts from those recordings. First among them was the owner of the statue and of the home in which it was displayed. Silvia Arevola was a mother of two in her mid forties.

Silvia: "I went to this shop. I saw that they had in the window statues that I did not like very much. So I went into the store and asked the owner if he had any other statues of the Virgin Mary. None of what he showed me I liked. The storekeeper then showed me a statue of the bust of Jesus of Limpias. I liked it immediately. It was love at first sight. So this is the one I bought and I brought it home."

Me: "What did you do with it after you purchased it?"

Silvia: "I brought it straight home."

Me: "Was it packaged in any way?"

Silvia: "The storekeeper wrapped it in newspaper and placed it in a plastic bag."

Me: "When did you unwrap it?"

Silvia: "As soon as I got home."

Me: "What did you do then?"

Silvia: "I started preparing a prayer room. I used to use this room as a gymnasium, so I had to move many things out. I brought other things in and moved them around. This took me all afternoon."

Me: "What happened then?"

Silvia: "At about 7 pm I asked my family to come and see the way I had set it up and to pray with me. My cousin was also staying with us. When she came in she asked if someone had been sprinkling water on the statue. Then she said this cannot be possible! We could all see the tears coming down the face."

I interviewed Oswaldo Rioja, a journalist from Channel 4TV who had been sent to file a report on what was causing so much attention in the local community. He set up a camera in front of the statue and let it roll.

Oswaldo: "At first," he said, "the eyes of the statue were a bit watery. After about 45 minutes a tear began to form in the corner of the eye and then to roll down the face."

He described collecting that tear on a piece of cloth. And then shortly thereafter, another.

Oswaldo: "We waited a while and then we saw on his left eye a tear started to roll down and I was holding the little cloth and when I went to wipe it off we realized and we saw that it was a drop of blood."

Me: "Was there any way that someone could have placed a substance on the statue without you noticing?"

Oswaldo: "When I came here I was sceptical about it. But I

was there for over an hour and a half and there is no way that anybody, nobody, even got close to it, and no way that anybody would have put water or any substance that would make it look like tears or blood."

Channel 4 was not the only organisation to film the phenomenon. On more than three separate occasions in 1995 alone, video cameras were set up to record, without interruption, the process of the formation of drops of blood on the forehead. Arising from apparently nothing but painted plaster they swelled and fell in rivulets down the face of the statue.

Of the many people I interviewed on videotape one was Catalina Rivas, also known as Katya Rivas. It was the first time we met.

Me: "You are related to Silvia, are you not?"

Katya: "My daughter, Tatiana, is Silvia's daughter-in-law. I love Silvia very much. There was a time in my life when I was like her; a girl very happy in the world. When I converted she was very ill, and she was almost dying. We prayed for her. She got well. We prayed for her conversion."

Me: "How long before she purchased the statue of Christ did you pray for her conversion?"

Katya: "About 8 months before."

She said that she had asked a priest to go and hear Silvia's confession but the priest had replied that he couldn't force her. The night before the baptism of their grandchild she had been praying the Rosary when she heard the voice of Jesus say, *'Be patient, she will confess.'*

The following day, 5 March 1995, both Katya and Silvia were at the baptism of their grandchild. They were both to be godmothers.

Katya: "The next day at the baptism Silvia went to confession and after many years went to communion. At the party she had a conversation with me. I don't know why I said this but I said, 'Give yourself totally to Our Lord,Silvia because I am sure this is a big sign for you.' Afterwards Silvia came to my house. She saw a small place where I say my daily prayers. She wanted to have a similar one in her home. She wanted to have a statue of the Virgin Mary. I gave her three or four addresses where she might able to buy one. She could not find one she liked. But she liked a statue of Christ."

Silvia did as Katya suggested and thus the statue was bought, brought into her home, and placed on the altar she had prepared. That very night it began to cry.

Me: "When was the first time you saw the statue cry?"

Katya: "The statue first cried on the 9th March 1995. My daughter told me about it. So the next day I came to see the statue and to pray. My mother was very ill and I wanted to pray for her."

Katya then went on to tell how she had obtained a piece of cotton wetted with the tears from the statue and applied it to her mother. Her mother recovered.

 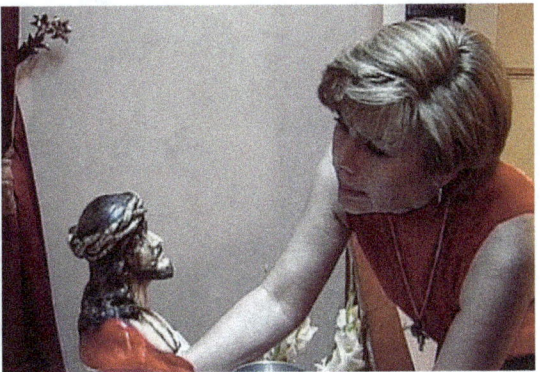

In Cochabamba Bolivia, Silvia Arevelo purchased a statue that looked like this. Six hours afterwards it began to shed tears. (9 March 1995)

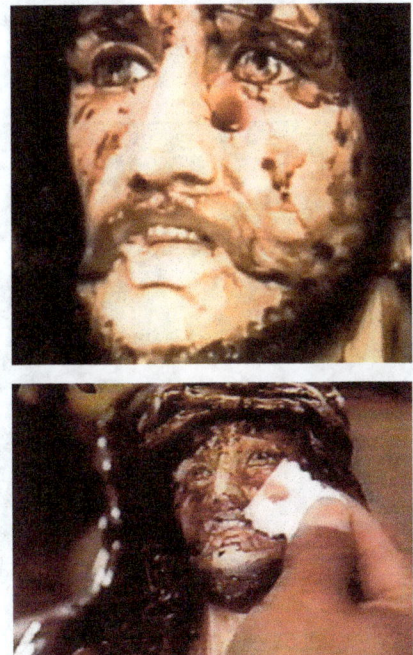

The statue was filmed shedding tears and blood. Samples were taken to test. (9 April 1995)

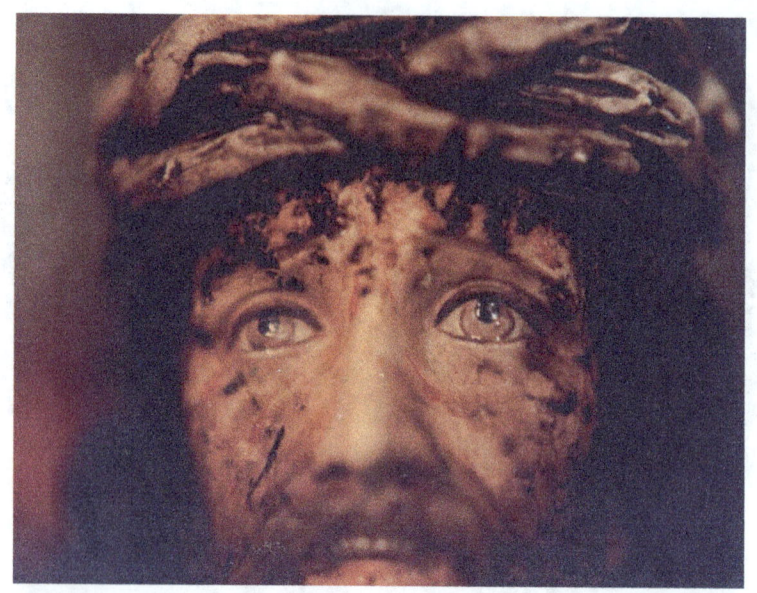

I filmed the statue shedding tears. (24 August 1995)

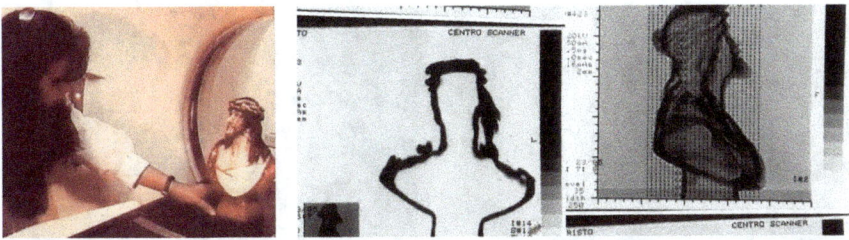

The day before I saw it cry, the statue was subjected to X-ray examination and was found not to have been tampered with.(23 August 1995)

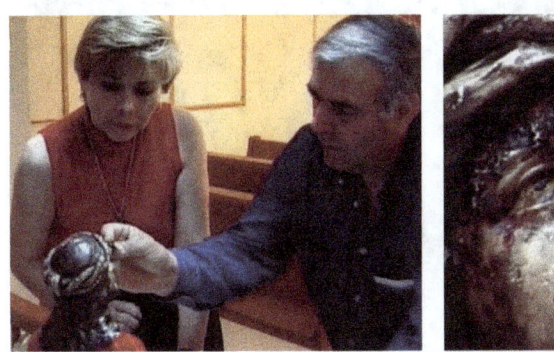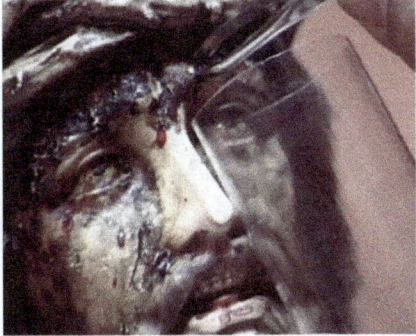

On 8 October 1997, I took a sample of blood encrustation and delivered it to Prof Angelo Fiori in the Department of Legal Medicine ,Gemelli Hospital, in Rome, for DNA analysis.

On 22 April 1998, Prof. Fiori reported that the blood was of human origin but had no genetic profile. He said," I have no explanation for this unusual phenomenon."

 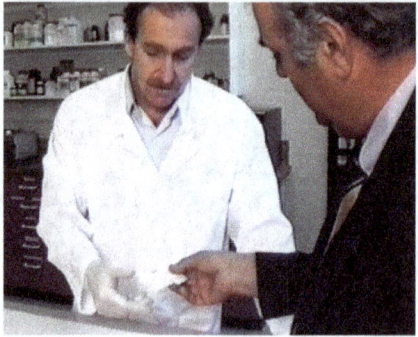

Testing of other samples I delivered to Prof. Sudhir Sinha of Gentest (USA 7.07.1995) and to an Australian crime laboratory (25.09.1995), produced the same result as Prof. Fiori.

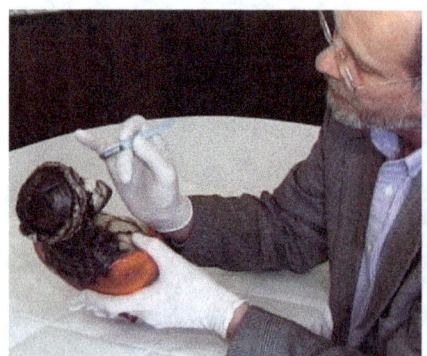 

Forensic Pathologist Dr Robert Lawrence went to Cochabamba and took samples for testing. He found that it was consistent with human blood.(31.1.2001)

 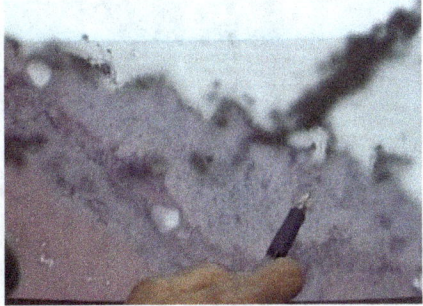

Under the scab was epithelial tissue (skin), reflecting the repair action of the body to stop bleeding.

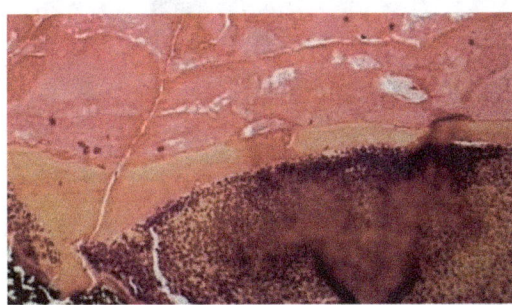 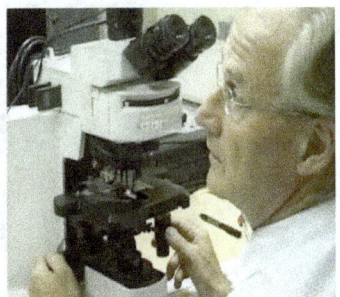

Dr Richard Haskell, an Australian Histopathologist and Cytopathologist, examined a section of material collected by Dr Lawrence. He found that there was injured tissue mixed with the blood. "The injury was likely caused by blows from a blunt object," he said.(14.02.2007)

 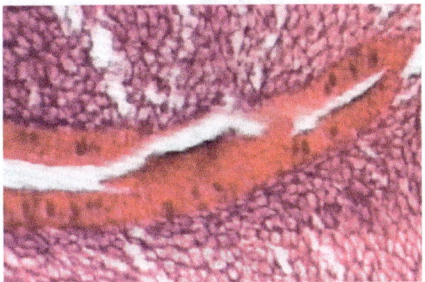

Dr John Walker from Sydney University found that embedded within a scab taken from the statue was the tip of a vicious thorn, likely from an arid region hawthorn bush. (February 1996)

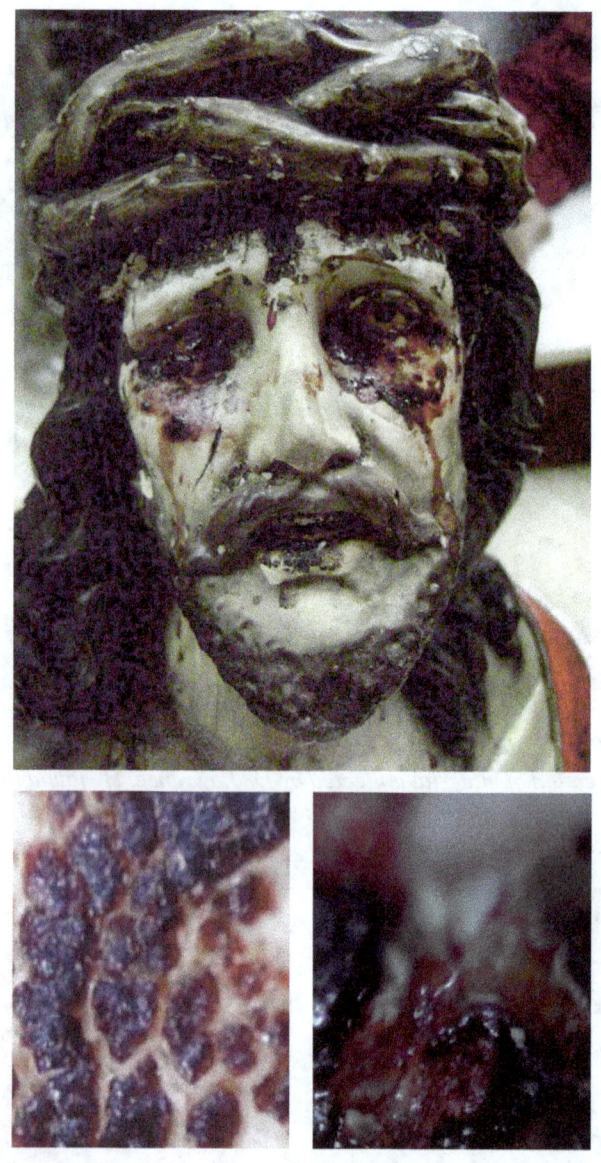

Samples were tested by Gene-Ex Laboratory in La Paz 'Bolivia. The samples passed the test for human blood, but there was no genetic profile.(14.02.2010).

CHAPTER 6
TESTS

Surely, I reasoned, the extraordinary phenomena which I had seen with my own eyes, filmed with my own cameras, and heard about from people who actually saw and experienced them, warranted scientific investigation. On 23 August 1995, with the owner's consent the statue was scanned.

The X-ray results showed that the statue was a solid plaster object moulded around an inner void. There was nothing but air in the void and nothing but plaster surrounding it. Had there been any perforation or crack in the plaster, any passage between the void and the surface of the statue, "even as fine as the diameter of a human hair," declared the radiologist, Dr Alberto Darras of Centro Scanners, Cochabamba, "it would have been detected."

Samples of liquid collected in gauze swabs by soaking up what appeared as droplets of blood and tears, (substances 'produced' by the statue,) and scrapings of the dried encrustations have been taken by me to laboratories around the world for scientific analysis involving DNA and forensic pathology. The reason for having samples tested in different laboratories by different scientists and different testing apparatus is to be absolutely sure that the results are scientifically reproducible. I have archived reports and filmed interviews by the scientists involved in the forensic analysis of samples taken from the statue. They have been consistent, and consistently intriguing.

Dr Sudhir K. Sinha, (Associate Professor of Biochemistry and Molecular Biology, who is well recognized in the international scientific community as being instrumental in the design of the first Y-STR DNA test for forensic use and as an expert witness on molecular biology and DNA analysis within the justice system of the United States) said, "When you look and see the colouration or the amount of red blood cells there, you would expect that there would be sufficient DNA in a sample to get a result. It is strange and difficult to explain that we can't get a result."

(GenTest Laboratories. New Orleans. USA.)

Professor Angelo Fiori tested two samples on two different occasions and reported, after testing the second sample, that "The new analysis performed on the blood stains allowed only to confirm that the examined material is blood and that it has human origin. However, and surprisingly, the new DNA analyses were again completely negative, although the specimen is quite abundant. I have no explanation for this unusual phenomenon."

(Department of Legal Medicine, Gemelli Hospital, Rome, Italy)

Dr Thomas Loy reported that by using standard methods for ancient DNA analysis developed in his laboratory, DNA was extracted and purified from the sample. These tests lead him to a conclusion that the blood was human but that blood was still surprisingly resistant to yielding a genetic profile.

(University of Queensland and the Australian Genome Research Facility, Brisbane, Australia)

Dr Susana Pinell Prado reported that the substance was deemed human blood, but that blood was still resistant to yielding a genetic profile.

(Gene-Ex Laboratory, La Paz, Bolivia)

Dr Robert Lawrence, Forensic Pathologist reported that the samples he tested in his forensic laboratory contained white blood cells with intact nucleii, fibrin, and red blood cells consistent with human cells.

There was something else. Dr Lawrence also described the sample as having a basal layer of epidermal cells. Skin. In an interview I filmed he said it looked as if "blood had been pulled from the face of a person like a scab that takes with it part of the skin. Delicate thin skin; skin that has no thick keratin layer, not skin that would come from the palms or soles of the feet or the back, but skin from the face."

(Delta Pathology and Associates, Stockton, California, USA)

Dr Lawrence was not able to identify certain clusters of cells present in the sample. These were able to be identified, however, by <u>Dr Richard Haskell</u>,

"There are clusters of cells throughout the blood sample. The cell clusters look like squamous stratified epithelial tissue. It is non-keratinized. It could be skin from the mouth, the inner parts of the lips or the inner parts of the nose. It is also possible that the tissue in the sample could have had the keratin layer stripped from it, and if that was so, the tissue would be from a deeper layer of skin from another part of the body such as the face. It would be abnormal to find this tissue in blood. So the admixture of fragments of tissue with the blood sample would seem to indicate that the material has come from some sort of wound, where the flesh has been ruptured and has produced bleeding. Such an injury would not have been caused by a sharp cutting object. A cutting object would sever the flesh and create bleeding without fragmenting the tissue. The injury more likely has been caused by a blow or blows from a blunt object which has fractured and traumatised the tissue admixing it with the blood of the wound."

Me: "Could someone have put this substance on the statue?"

Dr Haskell: "To do so the person would have had to use something like a serrated saw to cut their skin in tiny pieces and mix it with the blood and then place it on the statue."
(Australian Histopathologist and Cytopathologist, a specialist in identification of cells.)

Since my boyhood days of picking at scabs on my knees I hadn't given much thought to them. But once I had closely examined the statue and seen what clearly looked like scabs to me I was curious to understand how scientists understood them. So I approached an expert. A senior Australian surgeon, <u>Dr Colin Summerhays</u>, examined the slides prepared by forensic pathologists of the 'scabs' and the photographs of the 'scab formation' on the surface of the plaster statue. I interviewed him:

Me: Is a scab just dried blood?

Dr Summerhays: No

Me: So what is it?

Dr Summerhays: A scab is formed on skin after the surface layers of skin are damaged or breached. The first thing that happens is that bleeding occurs and this is followed by clot formation. Then the bleeding stops. At the same time, under the clot there's a complex series of reactions occurring and healing can commence. Epithelium starts to grow and to cover the surface of the wound.

Me: Would you comment on the scab pattern you see in this photo?

Dr Summerhays: It could have been an abrasion where there's a surface loss of skin. The slides from the pathologist of that scab show both clotted and degenerated blood and adherent to the underside is epithelial tissue.

Me: If I was to draw blood from my arm with a syringe, say, and put it on the skin of your arm, would it form a scab?

Dr Summerhays: No

Me: If one cut a dead body would the body stop the bleeding by forming a scab?

Dr Summerhays: No.

Me: A blood scab can only form on the skin of a living person?

Dr Summerhays: That is correct.

In transit, one of the samples taken from the forehead of the statue and placed in a test tube had broken into two pieces. What was originally a desiccated dark brown scab about 1cm in rough diameter was now in two smaller pieces. One piece was examined by two scientists:

Dr Robert Goetz, a specialist in DNA. He reported the same unusual result in that, "The sample passed the presumptive test for blood and was found to have the presence of human DNA. No genetic profile was obtained for unknown reasons."
(NSW State Forensic Laboratory, Sydney, Australia)

Dr John Walker, was puzzled. He looked at an unidentified cluster of cells in the blood sample from the Cochabamba statue and said, "They look weird. They look like the nuclei of liver cells."
(Forensic Pathologist, Westmead Hospital, Sydney.)

The second fragment of the original was examined by Dr Ian Clark, His report came back with information which went completely against everything every scientist had discovered. His half of a once single sample was not human in origin. It wasn't even animal. It was plant tissue. "It was", he reported, "tissue of non-human origin. The material has a definite structure, including nucleated cells. The material is undoubtedly not of human origin."

(Forensic Pathologist, Douglass Laboratories, Gosford, Australia)

At this point I was confused. How could a single sample, a single 'scab', break in half in a test tube and one half pass the presumptive test for human blood and the other half be plant tissue? To find answers I sent the mystery sample to Dr Peter Ellis, a specialist and lecturer in forensic medicine at the University of Sydney. He conferred with Dr John Walker who provided the answer. Like all the other scientists he had no foreknowledge of the origin of the sample. And like all the other scientists he permitted me to interview him on camera. The sample he was viewing under his microscope was projected onto a monitor which I could film while he pointed and explained what he was seeing.

Dr Walker: "It looks like the cross-section of a plant. Very small when you look at it with the naked eye. It's a matter of a millimetre or two in dimension. But in cross-section here, under the microscope, it has a mass of tissue here in the centre, this purple area, which is very small plant cells. Now given this structure, this very tight close structure and given the presence of these heavily lined passages, my feeling is that this is of a 'spine' which grows in a fairly dry sort of area. What it's designed to do is prevent as much fluid loss as possible from this plant tissue so that you have only got these openings to the outside from the lower side and this very waxy surface on the outside to prevent fluid loss."

Me: "When you talk about a 'spine' do you mean a thorn?"

Dr Walker: "Yes. Yes… The plant that comes to mind is a date palm which has really vicious thorns around the base. Or hawthorn. If they stick into you they break off and you get bits of them stuck in you."

The results from the tests so far show that the Cochabamba statue spontaneously produced human blood, blood with all the millions of particles, chemical and biological elements, molecules, organelles and tissue which subsist in normal human blood except for one very strange and unaccountable exception. No scientist can extract a DNA profile. The samples were examined in laboratories in five countries on four continents and not one was able to provide a simple DNA profile. The presence of DNA is repeatedly confirmed. It's there. It is in the cells. The quality and quantity of the DNA present is high. But it is completely resistant to being decoded. All nuclear DNA profiling has returned a negative result.

Repeated tests by scientists in different scientific institutions speak with one voice when they say there is no natural or scientific explanation as to why this statue has been found to have been exuding human blood. It would defy all the known laws of nature and biology for a plaster statue to spontaneously manufacture a single cell never mind complex human tissue. In this instance the statue is creating, ex nihilo, out of nothing, complete biological systems as complex as a human cell.

There is no explanation for the incorporation of skin under the scabs.

There is no explanation for the presence of traumatised skin cells in the blood.

There is no explanation for white blood cells in the blood. If someone had faked the statue bleeding by using their own or someone else's blood then a genetic profile would have been detected in every sample. That no human genetic profile could be obtained mystified many of the scientists.

There is no explanation for the presence of a broken thorn.

The world is home to many sculptures of greater aesthetic value than the little plaster sculpture in Cochabamba. Why, I asked, were sculptures of Winston Churchill, or George Washington, or Napoleon Bonaparte, or Buddha, or Ghandi, or Julius Caesar, or Shiva, not exhibiting 'behaviour' possible only to the living? Clearly something out of the ordinary was happening and only to the one representing a human being called Jesus.

CHAPTER 7
MESSAGES

'Look at your pen. For me you are my pen that I use to trace the symbols that express my words. What your hand writes, guided by mine, will remain. Repeated and amplified it will fill the earth.'

From *'The Crusade of Love'*, messages dictated to Katya Rivas

What I had seen and filmed, people I had interviewed - scientists, witnesses, journalists – and what had been revealed by scientific testing were all edited into a documentary I made called, "The Stones will Cry Out." In October 1996 I went to show it to an audience in Miami. On the first morning, while at breakfast in the hotel, I encountered Katya Rivas again, one of the women I had interviewed about the statue. She was writing in a notebook. She was, I was told, in the process of receiving dictated messages directly from Jesus.

Fascinated, and having my camera gear with me, I immediately started filming. I had no idea what to expect. I had no opinion on whether Jesus really was dictating to her or not. This is what happened:

She started on a blank page, wrote one sentence, then another, then a paragraph, then another, without pause or hesitation, without reflection, correction or alteration. She was completely calm and serene. She wrote in Spanish and never slowed to consider the form or content of the text. She never, unlike myself, seemed to search for a phrase or expression. It was as if she was operating on auto-pilot but was definitely not in any sort of ecstatic state. It was as if she wasn't deliberating, wasn't cogitating. She was fully aware of what was going on around her.

It may have been the first time a camera was able to record someone allegedly taking down dictation from a supernatural voice, but if assessed as authentic, certainly not the first time it had happened. The Catholic Church is entrusted with a veritable library of many such historical cases, some of which form significant corroboration and understanding of the Gospel message.

Later, after the writing stopped, I interviewed her.

Me: "Katya, you have been having some special experiences. Can you tell me about them?"

Katya: "Three and a half years ago I had an encounter with Jesus. The Virgin Mary first spoke to me and taught me how to say the Our Father. She was preparing me. She gave me messages and signs. Jesus asked me to get ready for a series of messages which will be his last call of love to humanity. I have already written four books of messages he has dictated to me."

Me: "How do you know that it was Jesus that was talking to you?"

Katya: "The first time Jesus spoke he said, 'Don't be afraid. I am Jesus. I want to give you messages for humanity. I want you to first learn how to say the Our Father. He started to lead me and I began to follow.

For the first time I began to understand the prayer."

Me: "Does he wait for you to finish writing before he continues?"

Katya: "He waits for me to finish the sentence."

Me: "Do you see him when he talks to you?"

Katya: "I see him with the eyes of my heart."

On later occasions I filmed her write responses to questions posed by people present. She had been asked to ask Jesus specific questions. She wrote complex, clear detailed responses without prior preparation. She could not have consulted anyone or referred to other books. Her writings were not the product of a photographic memory.

She had a rudimentary Grade 7 education, spoke only Spanish in a limited functional vocabulary, had no theological training and was of humble social standing. From her pen over eighty books have been written, a number of which have been published. Included in some are sophisticated passages in Greek, Latin and Polish, languages she cannot read, write or speak. Translators and literary experts have remarked on the beautifully lyrical use of language in the texts and of how difficult it is to adequately relay this in translation to English.

I have filmed Katya writing under dictation from Jesus on a number of occasions.

Intrigued, I sought some expertise on what were astounding claims by Katya. Father Omar Huesca from Miami had read and studied the messages compiled into books. What were his impressions?

Father Omar: "My impressions are that they are very much attuned with the message of the Gospel. There doesn't seem to be any sensationalism involved in the messages. They are very straight forward. They are a call to love, to commitment, in our every day life, which is something that impressed me very much, because they are implementable, they are messages that are reachable, that are able to be lived, and at the same time, great beauty, great depth of theology that impressed me. Impressed me very much."

Me: "You know the person, Katya, who doesn't know theology, and you are impressed by the theology. What do you say about that?"

Father Omar: "Well, I feel that if the messages were not dictated to her by a theologian, certainly they had to come from someone that is an expert theologian. In this case, we believe that it is the Lord Himself who is speaking to her. And so, I see no other explanation. Its either a total fraud or it is absolutely genuine."

Me: "If it were a fraud then presumably whoever drafted or told her what to write…"

Father Omar: "No, I don't think so. I think that if,.. it would be difficult for someone that does not have theological training to be able to memorise or to express these theological truths with the simplicity and directness which are evident in the writings."

Me: "Is there any one message that stands out, that you can think of?"

Father Omar: "Yes, the call to live a truthful life. I think that one of the great problems that we have in our day is that truthfulness is not something that we cling to in its fullness. It's very easy for us nowadays to justify telling an untruth or half of the truth to fit our needs of the moment and it seems to me like the messages that the Lord is giving Katya are calling for an absolute commitment to the truth. I had a Professor at University who told me that the source of all mental illness was untruth. It makes us depart into one degree or another of fantasy. And after many years, first as a psychologist and then as a priest, I can see this when I counsel people. When we begin to depart from the truth, we begin to enter into a reality made by ourselves, which can degenerate into mental illness, or at least, very often I feel, into the incapacity to trust others, and to be trusted by others. This emphasis on the truth is of the utmost importance to me, I think it is very timely and a great challenge to us."

Me: "If this is true, then God is speaking to man through Katya. This would seem to be an important epoch in time."

Father Omar: "I definitely agree, and it should not surprise us that especially at difficult times in history, God has chosen people, different people in different situations to stand or to communicate in a very specific way with an urgent message because situations have been so difficult that the solution has to be radical. What is curious is that He has always chosen the lowly so that the world may see that it is truly God who is calling us to this conversion basically, which is always the message – a message of conversion, of turning back to Him and receiving from Him the graces and gifts that we so sorely need, and that we long for, but that somehow we end up looking for in places that it cannot be found, where no satisfaction can really be found. So it wouldn't surprise me at all. I am a bit biased because I have met Katya, I have spoken to her, have asked certain questions, not to test her but rather to be able to gauge the scope of her experience as a mystic, and I have been deeply, deeply impressed by her; both in her human aspect, as well as what God is doing in her.

Me: "Do you think that the writings will be of value to the Church?"

Father Omar: "I think it would be of extreme value to the Church because they are again simple, to the point, but extremely profound and at a time when diversity in the Church is rampant. We need, I feel, to return to basics, to Gospel basics and this is what comes across so clearly in the messages. There is nothing new, there is no pretence of a new revelation, It is rather a new call to appropriate the revelation of God and to live it. This impresses me tremendously because I feel that there have been people who have either been imposters or have imagined having this type of relationship with God. And often those messages are sensationalist, sometimes morbid. The messages however that Katya relates in her books are extremely hopeful. A great deal of hope based on the immense love of God which is revealed constantly in the messages. A God of immense mercy, but at the same time a God of justice who will not violate our free will and will indeed allow us to reap the fruits of the seed that we plant in our lives, and so in a sense we are the ones that are called to make a decision, and I feel that it is very much needed in the Church today. I think we need clarity and we need commitment. That is what I see in these messages, a call to clear thinking, and committed action for the Kingdom of God."

Me: "Is there anything in the messages which relate to our immediate future?"

Father Omar: "I feel that there is, there is a certain urgency, a call to conversion. There is an urgency for us to take account of what we have done with the gifts of God. I feel that there is also an implicit message of what the results would be of our turning away from what God has commanded us to do, in our not giving the proper value to the gifts that He has given us. However, I haven't read in detail each page, but I don't see any specific predictions about doomsday happenings, or anything like, which is something that comforts me greatly because from the word of God and from the wisdom of the Church, we know that we are not to predict exact happenings, exact dates, etc. We are to discern the signs of the times and respond accordingly to these signs. I think there is a certain urgency there, but it is not a frightful urgency. It is a very hopeful one. Maybe someone else that reads them may find a different, may have a different impression, but that's basically what I have understood there and it has been very positive because it leads me to put more faith to be more – I find the messages to be more credible because of this balance, because of this very solid presentation, in accordance to the teaching and of the Church and of the Scriptures and everything else.

Me: "Some people would say the explanation for the experiences of Katya and others, is that it is not necessarily from God. It could be from the other side, the dark side. Are you able to say?"

Father Omar: "I don't believe so, because her experiences have always been for edification and conversion of not only herself, but of others, to Christ. I was saying before that it doesn't surprise me that this could be a charge that would be levelled against Katya because it has happened before, it happened with Christ Himself. Those who opposed Christ said that he expelled demons with the power of Satan, with the power of the head demon. Jesus retorted by saying how can Satan expel Satan, how can a house divided stand for long. I think that the adage from the Gospel, 'by the fruits you shall know them' is very very important, if not the most important tool of discernment. Now for the past several years the mystical experiences that Katya has had and the consequent messages have had the effect of bringing many people to the Lord, of converting people that once were enemies of the Lord. How would this be Satan? I really have a very difficult time accepting that whole proposition, or even thinking about it seriously."

In conclusion, as a psychologist and as a priest Father Omar said: "I think that from what I have seen and have heard from Katya herself and what I have read in the messages, and what has been told to me by others, it seems that this woman has received tremendous, tremendous and very unusual gifts. I believe that there are many mystics in the world that are anonymous that we do not hear about and I believe that perhaps many people, many good people, of constant prayer, serious prayer, do receive different gifts for different purposes at different times, but in a case like Katya's there is an abundance of manifestations that to me seem to be and I am lead to believe, because the manifestations are never to call attention to her, but rather to the message itself, and she is an extremely humble woman and from my point of view, as a psychologist, I feel that she is also a very healthy wholesome woman, as far as her emotional and psychological life is concerned, very normal, very down to earth, capable of enjoying a joke and capable also of being a bit upset when things are not the way they should be. When I see someone claiming to be a mystic or claiming to be holy, acting in a way that is totally dehumanised in the sense that they almost project perfection as one would see in the caricature of a saint, immediately I have a tendency to be a bit sceptical. Yet when I see the graces that are flowing through the life of this lady and at the same time see how human, how normal she is, I cannot help but be persuaded that this is very genuine."

Bishop Rene Fernandez of Cochabamba was privy to the mystical phenomena occurring in the life of his parishioner. Her early messages were given to him to read and assess. He declared them to be free of doctrinal or moral error and gave his seal of approval for them to carry the imprimatur of the Roman Catholic Church.

IMPRIMATUR

We have read Catalina's books and we are certain that their only objective is to lead us all on a journey of authentic spirituality that is based on the Gospel of Christ. The books highlight as well the special place the Blessed Virgin Mary, our Mother to whom we should offer our complete trust and love, as her children that we are, and our role model in how to love and follow Jesus Christ.

At the same time as they renew our love and devotion to the Holy Catholic Church, the books enlighten us on the actions that should characterize a truly committed Christian.

For these reasons, I authorize their printing and distribution, and recommend them as texts of meditation and spiritual orientation, so as to yield much fruit for Our Lord who is calling us to save many souls, showing them that He is a living God, full of love and mercy.

Mons. Rene Fernandez Apaza
Archbishop of Cochabama

2 April 1998

CHAPTER 8

SIGNS

It soon came to my attention that phenomena which were completely outside of what was normal were not only presenting in a plaster statue, they were also happening in human flesh, in a living person, to Katya herself. The dictations she received and the weeping bleeding statue were one thing. But quite another was that Katya said she had experienced, on multiple occasions, in her own body, the physical signs of the crucifixion of Jesus - the stigmata. During episodes of intense pain, bleeding wounds spontaneously opened on her hands, her feet and her face. The photographs she showed me of what had happened a few months prior were confronting; vividly discoloured contusions on her feet and palms, bruising on her cheek. This was entirely new and perplexing territory for me. What exactly was all this about? I began asking questions and filming interviews about her experiences of the stigmata of Jesus.

It first happened, she said, on 13th October 1994. She was kneeling in front of a large crucifix in prayer. She said she offered her life to the Lord (as a victim soul). It was then she began to feel pain in her hands and feet. She felt intense pain in her chest and fell to the ground. The pain went through her palms and her feet. After a few minutes she got up and saw that her hands and feet were extremely swollen.

The next time was the first Friday of January in 1996. She had noticed the night before that her hands and feet were swollen and yellow. Then on Friday, between three and four in the afternoon, they became darker, purple and she experienced a stinging burning sensation, like the pain, she said, of a real wound. In May 1996, after a similar episode, blood came pouring from her nose and mouth. In August her feet bled from spontaneous wounds, this time in the presence of doctors. The stigmata happened again on the 4 July 1997 in Tarija and again the event was filmed and witnessed by a medical doctor, Dr Tehran. I sought him out and interviewed him on camera.

Me: "Dr Tehran, you were present when Katya was having the stigmata, what was her condition when you saw her?"

Dr Tehran: "I was present about a year ago when Katya was having her stigmata. She had a clinical condition of someone who was in severe pain, and a condition which could lead her into shock."

Me: "Can you describe in more detail what you were seeing?"

Dr Tehran: "I would say from a medical point of view, she was in a state of severe pain just before the point of death."

Me: "What did you observe about the wounds she had?"

Dr Tehran: "I saw wounds in the palms of her hands and also some points on the forehead. On the forehead it looked like spines that were stuck into the forehead and they were bleeding fiercely."

Me: "Did the condition of the wounds change while you were there?"

Dr Tehran: "I did not see any physical change in the wounds, the only thing is that they were bleeding profusely."

Me: "How long were you present during the time of the stigmata?"

Dr Tehran: "I was there for approximately two hours. I was there as a doctor in the role of overseeing her, to see whether she would get better or that nothing would happen to her."

Me: "In that two hours, how would you describe the progression of what was happening?"

Dr Tehran: "On the physical and psychological side of it, it was difficult to see her because at times she was conscious and expressive, but at times she was also very vague, but the pain she was going through was like someone who was near death and as time went by she got better."

Me: "As a doctor, what do you say when you see something like this?"

Dr Tehran: "It is very difficult for me to give you a diagnosis of this because this patient was like someone who was near death – something that you cannot explain."

Me: "What were the wounds consistent with?"

Dr Tehran: "It was very interesting to see this because the wounds seemed to be made by something which had a very sharp point, as well as in the hands and in the feet and also there was like a dash in the palms of the hand, and this could have been done by a nail, in the shape of a nail and in the head it was little points in a smaller scale, as if they were made by thorns."

Me: "Is it possible that she could have inflicted these wounds on herself?"

Dr Tehran: "I believe no, because 24 hours later when I saw her and to my surprise the only thing she had were little dots on the hands and feet, and if she did that to herself, what is amazing is that 24 hours later they were healing."

Me: "Over what period of time would you normally have expected those wounds to heal?"

Dr Tehran: "It's impossible in 24 hours for them to be healed. I would expect at least 4 or 5 days to start to heal."

Me: "How long would you expect the wounds to take to heal to the point where you saw them 24 hours later?"

Dr Tehran: "I can't explain how it was done or why it was done, but as a doctor I can only certify what I had seen. It was healed in 24 hours."

Me: "As a doctor, what do you say when you see something like this?"

Dr Tehran: "As a doctor I have no explanation, but as a person, a believer in his faith, I believe in what I have seen. It is something Christ went through in the suffering on the cross."

Me: "Has this changed your life?"

Dr Tehran: "In a way it has. But really what it has done is reinforce my faith, which I have had for a long time."

Me: "When you discuss this matter with other doctor friends, what do they say about it?"

Dr Tehran: "When I have spoken to my colleagues I explain what I have seen, but they cannot explain – but what I can say is that these people do not make any fun of it. They do not say anything against it, and they say that this must be something extraordinary."

Extraordinary indeed.

CHAPTER 9

GIFTS

In 1998 I made a documentary about the unusual mystical experiences of Katya Rivas and was preparing to fly to Miami to present the film at a religious conference. She would be there too, with her spiritual advisor, a Salesian priest, Father Renzo Sessolo. Working through the night to get it finished on time found me frantically packing my bags to leave at the last minute. I was really exhausted. I prayed.

"Jesus, I am worn out trying to finish this documentary on time. I really, really don't feel like a trip to the other side of the world right now. With all that it takes to go all the way there and to be back in five days for work. But I will do it. For you. I hope you appreciate it."

Because I travelled with heavy camera gear I approached the task, as always, highly conscious of avoiding excess baggage weight. Only the absolute essentials were necessary. I pack with military precision. This time I was also taking books and tapes which made me even more scrupulous. I scrutinised the large new can of Gillette shaving crème that I had just bought specifically to take with me. Should I? Shouldn't I? Too big, I thought. I'm leaving it behind. Besides I would only need enough shaving cream for one or two shaves because I would only be away for three days. I'll just buy another one in Miami.

After the screening two days later in Miami, I casually mentioned to Katya that on the other side of the world in Australia it was the 11 May already. My birthday. She insisted on a celebration here in Miami and arranged an impromptu party for that evening. I was touched and surprised to arrive and find she had even brought me a gift wrapped with a bow. I opened it and fell silent. It was a 250 gram can of Gillette shaving cream exactly like the one I had deliberated over and discarded three days before. Same size. Same colour. Same brand. How on earth did she know that this ordinary humble item was the one thing I really needed right now? How did she know that directly after this little party I was planning to go out and buy this exact thing? There was nothing else to do and so, somewhat embarrassed, I asked her. She smiled and was reluctant to say anything. I asked again.

Katya: "I went to the shop because I wanted to buy you something. I didn't know what. So I asked Jesus. Please help me find something that Ron needs. I felt a tap on my shoulder so I turned around. There was no one there. But exactly facing me was a shelf of shaving crème. So I thought that must be what you need."

Father Renzo: "I was with her. I said to her no. I said shaving crème is not the proper thing to buy for a birthday present. It is too personal. Ron might be offended. But she said to me, Jesus wants me to buy it for Ron."

No one, not even my wife, not even the flies on the wall of my room back home, knew about my purely internal dialogue about whether or not to pack the shaving cream. It was such a small non-issue in my life it didn't even merit mentioning. Yet someone other than myself knew exactly what I was thinking and what I had planned a few days before. I believe that Jesus had heard my prayer back in Australia and wanted me to know he was appreciative of my effort on his behalf. I knew quite powerfully, that even though shaving cream is such a trivial thing, God listens to prayers and knows what we are doing and thinking. Even, and perhaps especially, the smallest details of our lives which to us may seem inconsequential.

It was in pondering the mystery of this memorable birthday gift that I found myself remembering other birthdays and that fateful birthday of mine thirty years before. And remembering the prayer I had prayed that day, the one I had not thought about at all for thirty years but which now struck me with the full weight of its providential gravity. I had prayed to meet the woman who would be my wife that night. And I did. Without her I would never be where I was, doing this work, thinking these thoughts and marvelling at the providence of God in the story he was writing with my own life.

CHAPTER 10

UNLESS I SEE

The world, I felt, really should know that the stigmata was a reality, factual in every way, more real probably than a fair amount of what passes for news in our media. I wanted to reach the biggest possible audience. I didn't want to tell people what was happening. I wanted them to see for themselves. With their own eyes. Had anything like it ever been seen publicly before? Not that I was aware of. If it could be filmed it would be the first time the world could see a human being spontaneously bleed from specific exclusive areas of the body for no discernibly material or diagnosable reason. But I was an amateur filmmaker. I needed help if I was to realise my somewhat grandiose ambition of sharing with the world what was going on. And in 1996, before the advent of the internet, there was only one way to do that: television. And as 'luck' would have it, help was not far away. I had no further to look than over my own garden fence.

Mike Willesee lived next door. To me he was a friend, a neighbour and a client. To Australia he was a famous TV star, a household name. Over thirty years his career as a journalist had brought his face to prime time television across the country. Renowned as a hard-hitting news reporter, he researched, produced, wrote, interviewed and travelled the world in pursuit of a story. He went after anyone anywhere to get the truth and would not back down. In the words of Channel 7, he was 'a good man to have on your side.' Every night his face was beamed across the nation as the creator, anchor and producer of a investigative program called, 'A Current Affair.' What he said, what he selected to show, who and what he considered newsworthy, reached millions of people. When he spoke people listened. More than that, they trusted him. In my situation of needing greater exposure it was like wanting a portrait of my wife and knowing that my neighbour was Leonardo da Vinci.

Being a rational, unimpressionable man with both feet always planted firmly on the facts of the matter, Mike was understandably sceptical at my revelations of what was happening in Bolivia. I wanted to show him the video evidence of the statue and of Katya Rivas' mystical experiences, but he refused to even look at it.

"Look Ron, these things can't happen and don't happen," he said with conviction.

"But it's all true though. And if we…. If you could just investigate it the way you do, better than anyone else, the way you do best, it would be the first time ever that the whole world would see it happening. Just imagine seeing a real stigmatic event in real time with your own eyes!"

"You mean go all the way to some unknown unpronounceable godforsaken place in the middle of Bolivia to see if some crazy person's hands bleed on cue?"

"Godforsaken?... Maybe not," I mumbled.

"It's all mind over matter," he insisted. " Or a kind of self-hypnosis. Or psychosomatic hysteria. Or probably a couple of religious fanatics needing some attention and getting up to some con tricks."

"If that's what it is then call it out. Expose it. Bust the lie." I cajoled. "There's a story in that surely?"

"Listen Ron, I've done stories like these before. Remember the one about the Filipino faith healers making a buck by pulling chicken guts out of some poor guy's abdomen and claiming they had removed his malady. Charlatans, the lot of them. Preying on the ignorant. Or that waterfall that was miraculously curing people in New Zealand? I'm sorry Ron. I've done this stuff before and it's always a big scam. And do you know what the first red flag is? None of them allow you to examine their claims closely. Like no cameras too close. No scientific tests or whatever else it takes and you can be absolutely sure that somewhere along the line there is always money involved."

His mind was set. But I wasn't giving up. I persisted.

"Look Mike. You hear me talking about all these things and without even a modicum of consideration you reject them out of hand. You're negative before you even know the truth. Worse. You've arrived at a conclusion without examining the evidence."

"Not negative. Just real. Been there, done that. Don't expect me to believe because you do."

"Okay then. Prove me wrong."

It wasn't long after that, that the 'good man to have on your side' was right by mine in a Boeing 747, over a vast expanse of Pacific Ocean flying to South America. True to form Mike had taken up the challenge. We were on our way to Cochabamba in Bolivia where Mike could assess for himself the strange phenomena of a statue spontaneously bleeding and where he could interview a woman who claimed to be receiving not only dictation from a man who had been dead for two thousand years but also experiencing the same physical wounds which led to his death.

CHAPTER 11

29 MILLION PEOPLE SEE

'Australian Story' was a popular documentary series on public television profiling Australians of note. Each week a video portrait was presented, each structured according to a set pattern whereby the subject directly tells his or her story. Shortly after our return from Bolivia, Mike was selected as the subject for an upcoming 'Australian Story' episode. Whilst researching archival footage for the production the producers came across the footage I had shot of Mike with Katya on our recent Bolivia trip. It was edited into the final cut and was broadcast nationally.

David Hill, a senior producer for Fox Television in the United States happened to be in Australia at the time and happened to watch it. He was fascinated by what he saw. He was intrigued that a journalist of Mike's standing would have independently funded research into a story which took him to the remotest parts of the globe, and to the remotest territory of investigative journalism. Anything to do with religion, with God, was routinely ignored and ridiculed. David Hill saw the potential for an investigative piece on what was happening in Bolivia. He contacted Mike. He commissioned a two hour television special hosted by Mike called 'Signs from God' and when it was finally broadcast in July 1999 the ratings proved David Hill's instincts correct. Twenty nine million viewers of prime time television tuned in to see what had never been filmed before.

The making of 'Signs from God' was itself full of signs from God. It became apparent from our research that there was a consistent timing associated with when rare cases of stigmata occurred in the past. Usually they occurred at the same time that Jesus historically experienced them, namely on Friday, or during Easter. With this plausible timetable in mind we planned the production, expecting the big event on Good Friday, a day which had brought a large production crew, loads of equipment, Mike and myself a long way and at great expense to Cochabamba, to the house of Katya Rivas. On the night before she received a message from Jesus. Oblivious to the intrusive scrutiny of cameras, lights, microphones and foreign strangers, Katya serenely wrote it down in her notebook seated at the kitchen table. When she was finished she laid down her pen and with calm assuredness addressed the camera in close-up.

Katya: "Jesus says it is not the right time. Tell Mike and Ron that. There will be no stigmata tomorrow. He says it will be in his time and you must learn to trust him."

Mike and I were crestfallen and by the looks on their faces it was obvious what the rest of the crew were thinking.

The next day, on Good Friday, there was another message. Again we were filming when Katya declared on camera:

Katya: "It will happen on the day after Corpus Christi (Body of Christ) and you will be able to come with witnesses and film what happens. I will have wounds which signify the Passion of Christ on my head, my feet and my hands. It will start around noon and end just after 3pm and it will heal by the next day. I know this will happen to me because Jesus said to me it will happen and I believe him and I trust him."

There was nothing we could do except pack up and fly thousands of very expensive miles back home.

Months later I was preparing to fly back with the crew to film the stigmata at the prophesied time and place when I became very unwell. For someone who had never been sick it was annoying and unusual. The doctor diagnosed a heart condition and strongly advised me not to travel. I was urged to have more tests. But I had waited months for this. I had to go.

At the airport check-in I was so weak I couldn't lift my own bags to be weighed. Mike had to do it for me. I couldn't carry my own cabin bag. I collapsed in my seat. By the time I decided definitively that I was too sick to proceed and wanted to exit the aircraft it was too late. The plane had started taxiing down the runway. I was really worried I wouldn't make it to Los Angeles. I was worried that my will hadn't been updated. I was worried that my family didn't have sufficient instructions about my affairs in the event of, what I believed was, my imminent mid Pacific death.

With difficulty I made it to Los Angeles and decided not to push my luck and to stay there instead of continuing on to South America. Repatriating my body from the United States to Australia, I reasoned, would be less problematic than if I continued on only to die in the mountains of remote central Bolivia. But then I realised that if I was to disrupt in even the slightest way the very tight schedule we were following to be able to be at the designated time at the designated place I would cause us to miss out on the opportunity to tell the world the most unique story I could ever imagine. After all, it's not every day God himself sets a time and a place and makes an appointment for the world to see what few had only ever read about. So I said to him:

Me: "Jesus. I am doing this for you. This travel. This work. You determine whether I live or die. I am going to continue on this journey. It's up to you."

I continued on my way. I caught all the connecting flights. After many long hours I couldn't walk unassisted from the plane to the airport terminal or my hotel room. I had made it in time for the feast of Corpus Christi but stayed in bed for almost all of it. By evening I was feeling well enough to join the film crew for coffee at Katya's house. After a while I noticed Katya became silent and distracted from the chatter around the kitchen table. She then got up and left. We found her later on her knees in front of a little shrine in her bedroom. She was praying and weeping in convulsive sobs.

It was beginning.

Whispering, Father Renzo said, "This is the mirror of our Lord's agony the night before he was crucified. He was sad unto death because he absorbed and experienced all the sin of every human being, past, present and future. Our sin was his inheritance and his suffering. For this he felt his own presence before the father almost unbearable. In his messages to Katya he says that it was not the scourging, not the nails and the thorns that were most painful. This was the hardest part….. Sighing… it is heartbreaking."

In the full glare of cameras and lights the intimate poignancy of this palpable sorrow was callously invaded. It felt grotesque, but no one, least of all Katya, insisted we stop. At noon the next day small puncture wounds formed on her forehead, one after another. We filmed as they began to bleed. A swelling on her left cheek became a dark purple abrasion as if she had been punched. On the white skin of her palms and feet small scratches appeared in the form of a cross, first on one palm then an identical one on the other. These then ruptured into open wounds, again symmetrical, again the same size and shape on both palms and both feet. They began to bleed. They penetrated the flesh of the hands and feet entirely, from one side to the other. All the while Katya was in severe pain, struggling to breathe, mumbling inchoate prayers and contracting in spasms as the hours passed. The wounds grew larger and more bloody. By 3pm I was extremely concerned. She seemed to be dying.

Cameras rolling continuously provided hours of footage, an incontestable alibi for the supernatural nature of her ordeal. The cameras were deliberately set to superimpose the actual time on the video images as they were recorded. The progression of events is unambiguous. The time the first changes began, the time they progressed, the time they began to heal is evident. The spontaneity of the formation and healing of wounds was indisputable.

Katya was not hypnotised. She was not pretending or performing. The wounds were not self inflicted. They were not created by makeup or magic or sharp hidden objects. There were no special effects. From the very first perceptible physical changes on her skin to wounds which pierced her hands and feet all the way through issuing copious amounts of blood the cameras were rolling. I saw it. Other witnesses in the room saw it. The camera crew and producers saw it. And at 8 pm on 29 July 1999 so did twenty nine million people in the United States and Canada.

The next morning we were greeted by a cheerful smiling Katya, her old self in every way. No evidence of injury was visible. I wasn't sure what was more amazing: the mysterious appearance of gaping bleeding wounds or their astonishing disappearance within eighteen hours without leaving so much as a trace of ever having been there. No bruising. No scars. No sign of trauma at all. The crew was perplexed but Katya was confident and peaceful when she declared that Jesus had come with a message at the end of her ordeal.

"I have been preparing you for this day because I need to reach the world one more time to show the world my suffering through a person like you. Thank you."

There was more; a message for me:

"Tell Ron there is nothing wrong with him. What he must do is breathing exercises, take aspirin, rest."

Mike took me outside and on the footpath of a public street taught me how to do breathing exercises. I began to feel better immediately. I took the aspirin. I rested. I recovered fully and I did not see a doctor, not for my heart, not for any reason whatsoever, for the next ten years.

On Good Friday 1999, there was an unprecedented prediction of a supernatural event. Before 3 cameras and 5 witnesses besides myself, Katya said " Jesus said I will have the stigmata on the day after the feast of Corpus Christi. It may be filmed for all to see."

'Signs From God' was broadcast to an estimated 29 million people (29 July 1999).

As predicted, on 4 June, the wounds of Christ's crucifixion began to appear on Katya's body and were professionally filmed continuously with embedded timecodes.

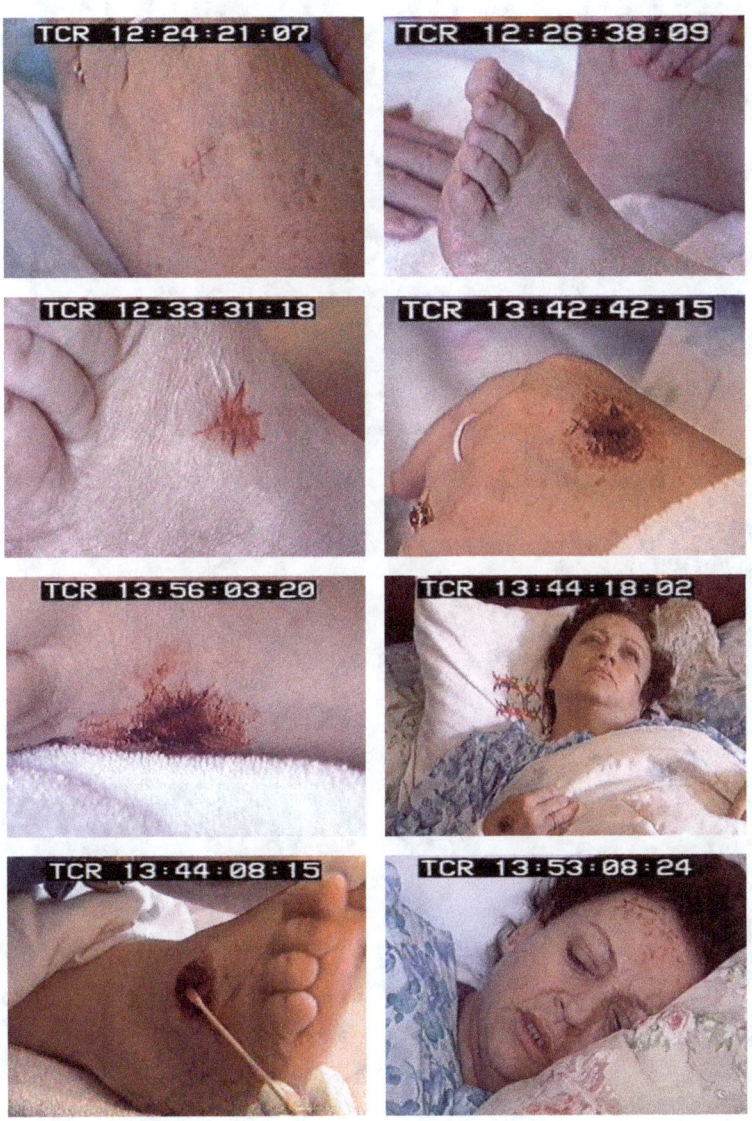

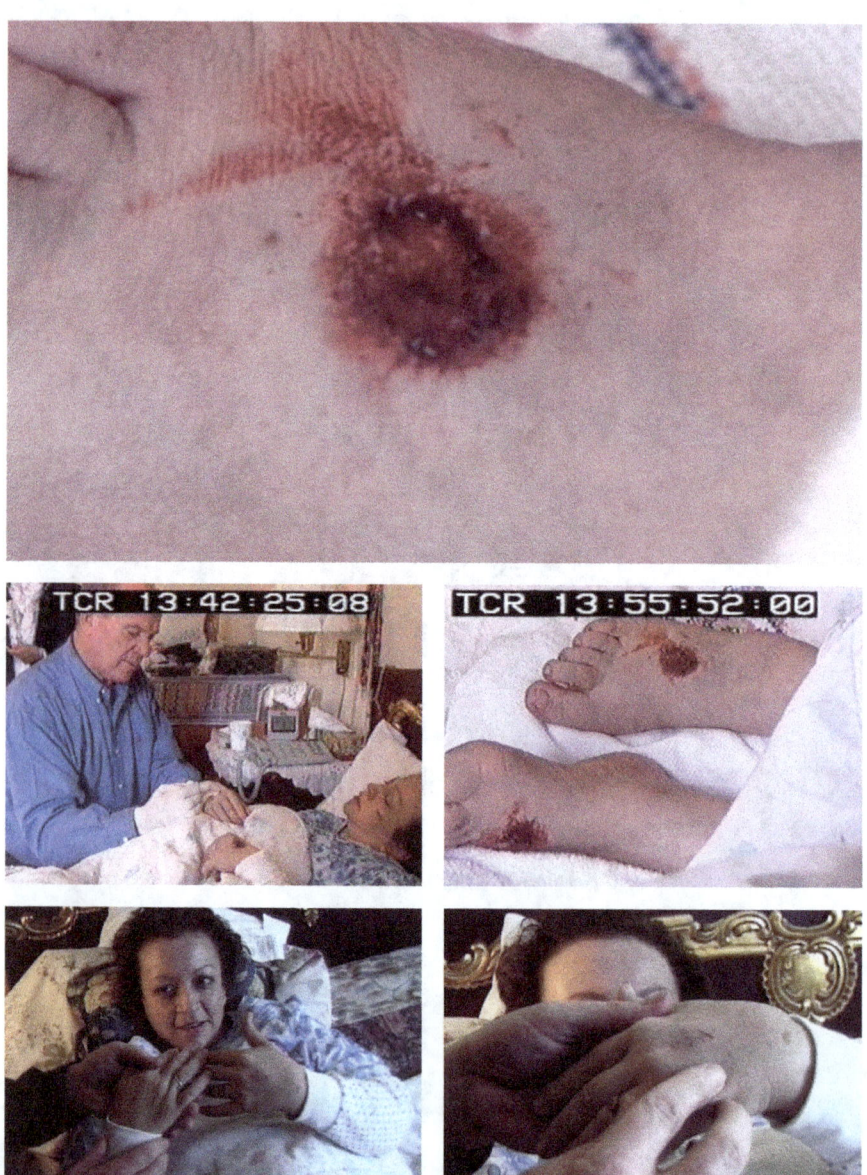

By the following morning the wounds had completely healed contrary to medical expectations.

CHAPTER 12

UNLESS I PUT MY FINGER

A few months after the world saw the stigmata I interviewed Mike Willesee.

Me: "Mike, you were there when Katya had the stigmata. What was it like to be there and see it for yourself?

Mike: "It was actually a frightening experience. I had observed this woman for some time; from the position of me sitting on the fence in the first place, to finding that she was a humble, holy woman, not making money, not trying to form a cult and I was really impressed with her. But like a lot of us, I think, well, what if the stigmata is not real, what if it is a deceit? And I'm there with her, a crew, and I expose it? If I found deceit there I would have to expose it. I really didn't want to because Katya was winning me over. What we saw was the exact opposite. It was credible. We were there for more than the three hours of the stigmata. There were four of us. We video taped hour after hour, and I simply could not fault it. I can't see one possibility of how there could have been fault."

Me: "Mike, people might say, that if Katya had the wounds that you were able to show, that either she did it herself or someone did it to her. They just can't happen by themselves. What do you say about that?"

Mike: "Having been there I don't believe it was self inflicted or put there by somebody else. It started on the back of her left hand in the form of a cross. Now that in itself was astounding to me, because I couldn't determine whether that cross was put there by a drawing, by a pen, by ink, or by a cut. It didn't seem natural. I don't know how you can cut a perfect cross in skin without causing bleeding. It was a perfect little cross which as you saw we filmed. That then became a wound, an open wound. A matching wound, and I stress, a matching wound appeared on the other hand. Matching wounds appeared on the top of her feet, and on the bottom of her feet. I know those wounds were true and real, because I put swabs into them. I actually interfered with those wounds, I took blood from them. As I say, they were matching. We were there all the time. We literally saw the first wounds on her forehead appear. There was a spot of blood, and then a second, and then a third spot of blood. She is lying there, she didn't touch herself. It is not possible that she did that to herself and if somehow, beyond all our belief, or all our credulity, she did do that to herself, when we went back the next morning, the forehead was clean, no wounds. The hands were clean, no wounds, but now two crosses, you would not describe them as wounds. Two crosses on the back of the left hand. On the top of the feet you could see the marks, but when you felt them, they were completely smooth to the touch. Now, I don't believe any microsurgeon in the world could heal those wounds so smoothly in less than 24 hours."

Me: "Mike, Katya predicted that she would have the stigmata on a certain day, and it happened on that day, and you were there to film when it did happen. What do you say about that prediction?"

Mike: "We first went there on Good Friday believing that Katya would have the stigmata on Good Friday. There were a lot of reasons. We weren't told that she would, but there were a lot of reasons for us to presume that she would. She had been having stigmata on Fridays, and Good Friday was the day of Christ's crucifixion. We expected it. We expected it too much, and it didn't happen. Katya said she had a message from Jesus that afternoon and said don't be disappointed, have faith. Come back the day after the Feast of Corpus Christi. That is the Feast of His Body and Blood. That was two month's away. And we did. The Fox Network that we were dealing with said, 'Can you really believe that? We are starting to have doubts' and I said, 'Well everything that we have been told so far when Katya has said she has a message from Jesus, has come true. And I'll certainly be flying all the way from Sydney to Cochabamba to be there on that day.' We did, and at 12 o'clock, as predicted, on that day, the stigmata started. The very fact that she predicted the date two months away, and the time, and it happened, lends a lot of credibility to what we saw and what we now believe."

Me: "Mike, Katya says she has written many books that Jesus has dictated to her. How do you satisfy yourself that she is telling the truth, and these are from God?"

Mike: "I have very little experience in religion or theology, but Archbishop Fernandez of Cochabamba has given the imprimatur to this so he has had his Commission of Theologians examine all these works and he has found, and his Commission has found, that they are completely consistent with the Bible and the Gospels. I accept that, and from my limited reading of them, it certainly seems to be so.

In addition she writes in languages which she says she has no knowledge of, and which we can't find any indication that she might have knowledge of. And she writes those languages grammatically correctly. Now that takes a lot of training, it takes time being in the countries where she hasn't been. It is hard to believe that you could sit in Bolivia and learn a language so well. It is astounding when you think that this person writes free-flowing non-stop, page after page, book after book, and doesn't make a mistake, and I would defy an expert in any field to do that.

While I would regard her as an intelligent woman, this is a woman with very little education. I certainly don't believe she has the capacity to do that, nor do I believe I have ever met any person with a capacity to do that. I am completely convinced by what she writes."

Me: "Mike, you have put your professionalism as a journalist on the line in telling this story. Has it been worthwhile?"

Mike: "Yes it has. It has certainly been worthwhile for me. It's been worthwhile in the reactions we have had. I think about 23 million Americans have viewed the program, and millions of Canadians, Australians and New Zealanders, and I think it will go around the rest of the world, there is a great demand. The personal reaction I have had to it has been the strongest from everything I have done in all those decades. Some people have been converted by it. Some people have said, "Well I have got to stop and think". Some people still reject it, and say "Oh I saw your program," and then stop, then, "I still don't believe that, I can't believe that". But at least they are thinking about it, and if the message of Jesus to Katya is true, that man is turning his back on God, that man of the 20th Century has forgotten God, then at least people will have had reason to think about it.

So I feel it was very much worthwhile and I must say that after I went through my period of scepticism and had a number of experiences, I certainly came to the belief that the hand of God was visible in what we were doing. What we were told did happen."

Me: "Mike, if we believe Katya, God was using your programme to deliver a message to the world. Do you believe that is the case and what do you think is the message?"

Mike: "Katya said that Jesus said to her after that stigmata, 'Everything you have done,' which I think you could interpret as 'everything I have done with you,' her previous stigmata, all her other experiences, 'was in preparation for this day.'

Now if that is true, Jesus was saying to the world, remember what I did for you. This was my crucifixion. I was the Son of God. I didn't have to go through this, but I subjected myself to this humility, and this pain, to save you, and you have forgotten that, and I am showing you this pain again through Katya."

Nailing a thirty three year old Nazarene Jew to a cross in Jerusalem two thousand years ago was not newsworthy. It should have passed into obscurity alongside the millions of other gruesome deaths before and since. The victim, an unusual rabbi, had a small following while he was alive, but as he was being publicly tortured to death only 8 percent of his closest companions stood by him. Almost his entire inner circle, including the man he personally elected as their leader, abandoned him. All his teaching, all his promises, all his healings and all his miracles were not enough to drive out their fear of being associated with him. They did not believe that he was who he said he was: God.

But all that changed three days later when he rose from the dead and was seen alive, walking, talking and even eating. His apostles were overjoyed. His disciples were elated. But two of them were not around. One, Judas Iscariot, hanged himself in self disgust and despair for his part in betraying Jesus to his executioners. The second, Thomas, was the last apostle to make his way back to the group. The others all told him they had seen their dead master alive. But Thomas was sceptical of their exuberant claims. Thomas was adamant. He would not, in the absence of personal material first hand evidence, believe that a man who was declared dead and buried on Friday was walking around three days later. "Unless," he said," I see the nail marks and put my hand into his side I will not believe." Thomas had to actually touch an actual heart before he would believe.

This had been Mike Willesee's sentiment exactly. Like Thomas he doubted at first. And like Thomas he experienced an event, the stigmata of Katya Rivas, which was so extraordinary and so far off the scale of 'normal', that he came away a fundamentally changed man. Like Thomas, he fundamentally shifted his attitude after a very real experience of an inexplicable supernatural power over nature.

Jesus did not rebuke or disown Thomas for his lack of belief. He understood Thomas and his scepticism. His response was to patiently address his doubts with tangible material evidence. He invited him to touch the wounds created by his crucifixion. He directed him to touch his heart, his actual heart.

Mike too came up very close to the selfsame wounds. Like Thomas he put his finger on them and like Thomas, became a convinced and convincing witness completely overwhelmed by the evidence and able to cry out, "My Lord and my God."

In his memoir, 'A Sceptic's Search for Meaning', Mike described his life at the pinnacle of his celebrity career in journalism reporting on world leaders, wars, the great, the rich and the famous. He then made an astonishing declaration: Nothing came close to being as relevant to him as the Katya Rivas story. He wrote in my personal autographed copy: 'Ron, how can I thank you for leading me to my biggest story?'

As has happened so often in this work, all superlatives inevitably fail. So when Mike wrote that the Katya Rivas story was the 'biggest' of his stellar career he was wrong. A much bigger story was waiting to be told. I interviewed Mike on camera a few weeks before he died. We reflected on the experiences we had shared.

Me: "As a journalist whose duty it is to report on what you uncover, on the evidence. How difficult has it been for you, as a journalist, to tell the story knowing that most people won't believe it, even though you say it happened?"

Mike: "That doesn't bother me. Especially as I have filmed evidence, video evidence, which in America that night, or in North America that night, 28 million people saw that program go live to air and that is a big number in North America. You have to go to the Super Bowl or something to get something bigger. Of all those people, nobody came up with a genuine criticism. In fact there were very few criticisms because they didn't have to listen to me telling the story. They saw it happen."

Me: "If this story is true, as you say it is, you have a situation where a person has predicted the happening of it, following a statement they say Jesus made to them, that he will have the stigmata appear on the person's body and that you will be able to film it on a certain day. As a journalist reflecting on that fact, you're talking about a person who died 2000 years ago, talking to someone today, about the injuries that were inflicted upon him to cause his death. That's pretty remarkable stuff, isn't it?"

Mike: "Not just talking about it, but showing it. And this was very difficult for me as a non-believer. And I couldn't rationalise it in my mind. I knew what I'd seen. I knew what I'd seen was authentic but to me where was God? Was there a God? My confidence started to shake. No longer could I say to you, I don't believe in God. I would have to say to you, I don't know if there's a God. I've seen some powerful evidence that there is, but I don't know how authentic it is to say it came from God. It was a difficult period for me personally, because I had to make my own decisions. I wasn't just working for all the people who would see the program. I was working for me, too. And probably I was the most important person in the audience, for me. Is there a God? Is…There… A… God?"

Me: "Reflecting back now after 20 years, have you refined your assessment on that question?"

Mike: " I now have no doubt there is a God. There are so many other things that happened that were not in the natural sphere of things. Supernatural – the word that I didn't like at all. I saw so much of it. I witnessed prophecies, I witnessed events. There is NO doubt in my mind that God exists. It just took me a long time to come around to accept that."

CHAPTER 13

BUENOS AIRES CASE

In August 1996, in a busy commercial district in Buenos Aires, a small round wafer the size of fifty cent piece, thin and dry like a biscuit was placed in ordinary tap water in a glass dish and locked in a dark sealed metal container much like a wall safe. The wafer was an ordinary common baked product and no different from billions just like it which had been made exactly the same way for thousands of years. It was manufactured by mixing only two ingredients, refined wheat flour and water. After a week in water the wafer was found to have undergone an unexpected transformation. It appeared to be exuding a bright red substance which, to the naked eye, resembled fresh blood. The authorities were informed and a professional photographer was commissioned to document the transformation.

The glass dish and its contents were locked away in a new, secure and undisturbed location. Remarkable changes to the original wafer continued over the next few weeks. More of the red substance appeared, as did darker gelatinous patches resembling coagulated blood which seemed to be replacing what had previously been the white wafer. The photographer was recalled and the resulting photographs confirmed that the phenomenon was one of continuing transformation. About a month after the initial discovery the contents of the glass dish, the autonomously transforming wafer and the original water, were transferred to a sterilised test tube containing distilled water.

For three years the test tube was locked away. Then it was re-examined by eight scientists from three continents. The results were inexplicable. The substance which had spontaneously appeared in the wafer was found to contain human DNA. It was infiltrated with white blood cells. It was described as muscle tissue from the left ventricular wall proximate to the valvular area of a human heart.

Unleavened bread placed in water had become, within days, complex living tissue from a human being, the most complex species on earth.

I am privileged to be in a unique position to address the questions and the implications of this case because I had been called on to visit the Santa Maria Catholic Church in Buenos Aires on 5 October 1999 in order to examine and document what had occurred. I was physically present and able to interview on film, prime witnesses who saw the material before, during and after its transformation, all of whom I found credible and unmotivated by self interest. Furthermore, I personally engaged the many scientists who have been involved in this investigation over many years. I also spoke to and recorded statements from these very scientists.

According to all involved there seemed to be nothing extraordinary about 18 August 1996. They were present at the 7pm Mass being offered by Padre Alejandro Pezet. Assisting the priest in the distribution of communion was Emma Fernandez. Near the end of the distribution a parishioner approached Signora Fernandez and alerted her to a matter of grave concern to Catholics. She had seen someone abandon a consecrated host in a nondescript bronze candlestick, one of a pair next to a crucifix on the far side of the church. Signora Fernandez followed the informant to the place she indicated. Besides being dirty and dusty, the host abandoned there was no different from the thousands of others used every day, month after month, year after year. Signora Fernandez informed the priest, Padre Alejandro Pezet, of what had transpired.

In an interview I filmed he said,

"When I had finished distributing Holy Communion to the congregation a lady came up to me and said she had seen a communion host abandoned in the church and that it was in a candle holder at the base of a crucifix at the right hand side of the church.

I followed the lady to where I saw the abandoned communion host. It was in a candle holder we do not use very much. It was very dirty. I picked up the host and took it to the altar with the intention of consuming it. I saw that the host was very dirty and so I asked Emma Fernandez to put it in a bowl of water and put it in the tabernacle. She immediately went ahead and did so. I was very much concerned that the host was abandoned by someone who had performed an act of profanation with it."

The instructions Padre Pezet gave Emma Fernandez about placing the host in water to dissolve were compliant with long established practices outlined in liturgical guidelines. Since every particle of the consecrated host is believed by the Catholic Church to be God himself substantially, really and truly present, a consecrated host is the most sacred material object on earth. Not to treat it as such is considered a serious transgression. Every particle of the host is to be solemnly, entirely and immediately consumed by the communicant. Priests are to ensure that every last drop of wine and crumb of bread is consumed or, as in this case, where the host was too dirty, that it be entirely dissolved in water and then properly disposed of.

Proper disposal entails dissolving the host in water so that it is no longer a recognizable host and then pouring it into a sacrarium, a special sink from which the water flows to consecrated ground within the church precinct. The altar cloths and chalices used in the Mass are all laundered and washed in this sink so that any crumbs or drops of wine inadvertently remaining on them may be disposed of with due reverence.

Emma Fernandez described in a video interview how she filled a glass bowl with tap water from the presbytery, placed the host in the bowl so that it might dissolve and then placed it right at the back of the tabernacle. Some eight days later, on the morning of the 26 August, she unlocked the tabernacle intending to check on the bowl to see if the host had completely dissolved yet so that she could properly dispose of it. The bowl was in the same position in which she had left it. The contents however had changed considerably. A red substance was now surrounding the host. Alarmed by the radical transformation, she immediately locked the tabernacle and went to report her finding. Within minutes she brought Alejandro Pezet and another resident priest, Padre Eduardo Graham to see. Together they witnessed the same phenomenon when she unlocked the tabernacle and showed them the bowl.

Padre Pezet confirmed that, "On the morning of Monday the 26 August, I came down to pray in the Blessed Sacrament Chapel of the Church. Emma Fernandez found me there and said, 'Padre Alejandro, this morning I looked in the tabernacle and noticed something strange.' I got up and went to see. I saw that the host in the tabernacle was becoming red. This had such an impact on me. I felt that what I had seen was something supernatural."

Eduardo Graham, the other parish priest, confirmed the events of the morning of 26th August 1996. He too saw that the host was transformed.

It goes without saying that the properties of bread are not such that when placed in water for a week one would expect a bright red substance to come from it. So later that day, Marcelo Antonini, a professional photographer was engaged to visually document the phenomenon. From a bird's eye view he photographed a round glass dish with the white host still largely undissolved with its original circular shape. The circumference however was no longer crisp and showed signs of dissolution. Dark spots surrounded by bright red liquid appeared to be thoroughly integral with the host. The relationship between the original host and the new red material did not, to the naked eye, appear to be some foreign matter added to the contents of the glass. Instead it appeared to emerge from within the host.

When I interviewed the photographer about taking the photographs he said, "It was very impressive. There are no words to explain what happened." Nearly twenty years after this first interview he was asked again how he felt taking those first photographs. He appeared overcome with emotion. His voice faltered and he choked up.

Padre Pezet moved the bowl from the public tabernacle in the church to another private tabernacle in the presbytery. He locked it. Access to this key was available only to the three priests: Pezet, Graham and Juan Carlomagna, who was away at the time. "I saw it," said Pezet, "when it was first put in the tabernacle and locked. After we noticed what had happened I moved the bowl containing the host to the tabernacle in the presbytery."

To the astonishment of Pezet, "This bloodlike substance coming from the host grew in quantity over the following weeks," an observation echoed by Graham in his testimony. The fact that it continued to transform prompted the recall of the photographer to document the changes. By the 6 September the photos showed that the dark sections had grown larger and even more red liquid was interspersed with liquid of a ruddy brown colour. The original perimeter of the host remained only faintly discernible. Clearly some process of growth and change had been spontaneously occurring. A month later the transformed host was transferred to a glass test tube with distilled water, and it was sealed. It remained in the care of the Santa Maria Church on the Avenue la Planta for more than three years.

Then circumstances within the diocese changed. The ailing Archbishop Quarracino of Buenos Aires died. He was replaced by Archbishop Jorge Bergoglio, now known internationally as Pope Francis. Bergoglio considered it prudent to appoint professionals to examine the phenomenon which had occurred in his jurisdiction.

On 5 October 1999 a sample of the material was excised and placed in a sterilised test tube. I was present and filmed the entire procedure of taking the sample, sealing it and labelling it. This contributed to maintaining a clear, unbroken and recorded chain of custody. This I believed was essential in the investigation. I then flew with the sample to deliver it to a genetic testing laboratory, Forensic Analytical in San Francisco. It was referenced simply as 'Case 19990441/1/2.'

On 18 August 1996 Padre Alezandro Pezet found an abandoned Communion host in his church after Mass. He placed it in a bowl of water and locked it in the tabernacle.

On 26 August 1996, when the tabernacle was opened, he noticed that a red substance was coming from the host.

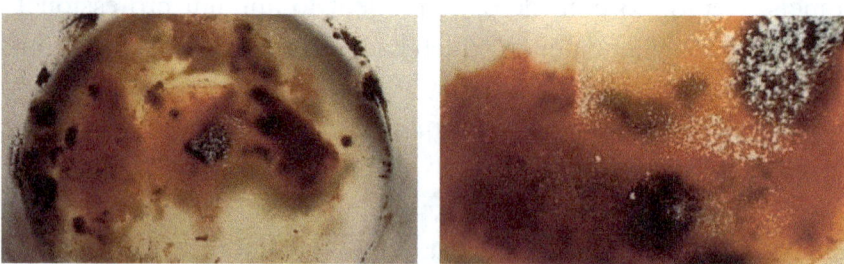

On 6 September 1996, when locked away in another tabernacle, the host continued to transform.

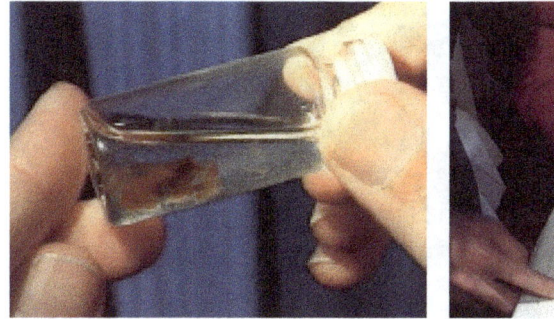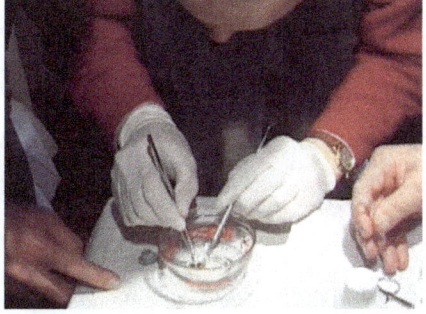

On 5 October 1999, I was asked to film and document the scientific examination and was to take on the principal role of leading the investigation.

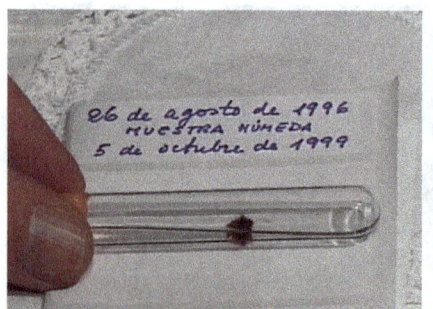 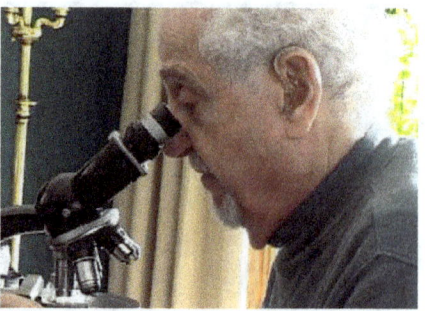

On 20 April 2004, I engaged New York Forensic Pathologist and Cardiologist Dr Zugibe to examine a sample of the substance without telling him its origin . Looking into his microscope and in the presence of only Mike Willesee and myself, he said on camera, "This is **heart muscle tissue**. It is infiltrated with white blood cells. **This person has suffered traumatic injury** like I have seen in cases where someone has been beaten severely over the chest. **This tissue is from a living person , not a dead person** ."

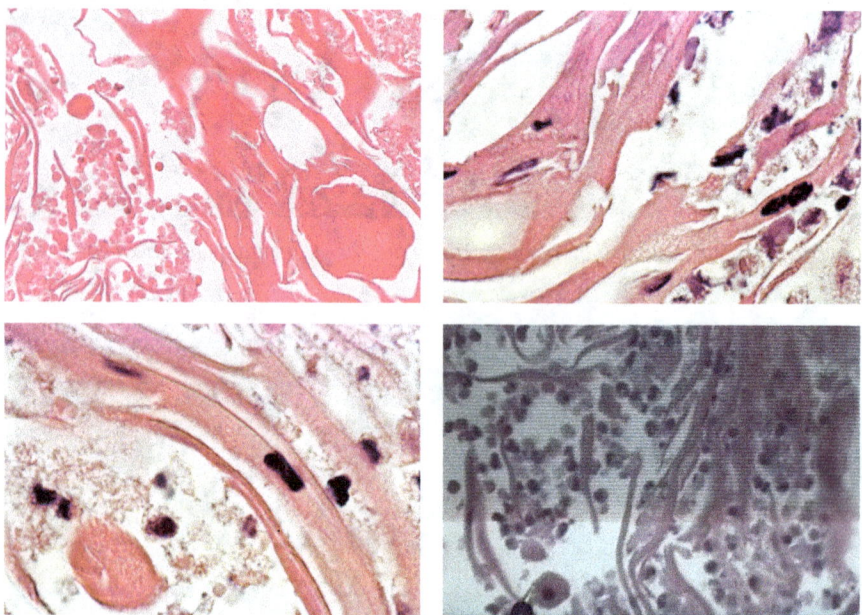

Microscopic sections of sample revealing inflamed heart tissue and white blood cells

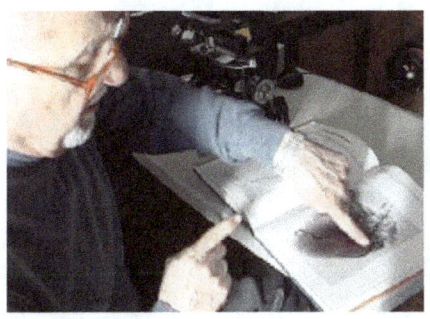 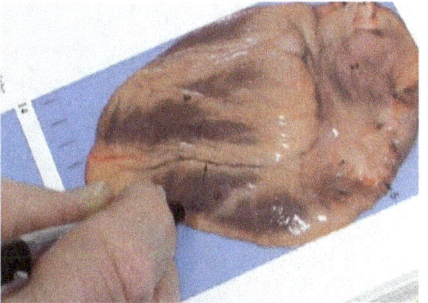

Dr Zugibe showed us the position from where the heart muscle tissue had come close to the left ventricle wall. (20.04.2004)

I was present when my report of Dr Zugibe's findings was handed to Cardinal Jorge Bergoglio of Buenos Aires. (17.03.2006) (Pope Francis)

CHAPTER 14

CASE 19990441/1/2

*The heart is the very first organ to develop
in the human embryo. Twenty one days after conception
a group of progenitor cells suddenly begin to beat,
even though there's no blood
to pump or vessels to pump it through.
It develops into the most robust
efficient machine in the world.
It never grows tired and seldom misses a beat.*

The results from these tests were remarkable. They revealed the presence of human DNA. Strangely though, the scientists were unable to extract a DNA profile.[3] On 9 March 2000, based on the recommendation of Forensic Analytical, I authorised for the sample to be sent directly to the forensic pathologist, <u>Dr Robert Lawrence</u>. He prepared microscopic slides from the material and examined the results. The sample, he said, contained epidermis (outermost layer of skin) infiltrated with white blood cells. In his report he described 'an aggregate of keratotic and parakeratotic debris with enmeshed leukocytes. Also there are scattered minute aggregates of brown material composed of septate hyphal fragments enmeshed in proteinaceous matrix.'

During a filmed interview on 7 December 2000 he commented on the presence of white blood cells saying, "They were active living white cells at the time they were collected. There was some inflammatory process going on." He was emphatic that, "if this material had been placed directly into water after it had been taken off a body, I would expect these cells to be dissolved in minutes to an hour or two."

I subsequently arranged for the microscopic slides to be examined by Australian scientists, Dr Peter Ellis, a senior lecturer in Forensic Medicine at Sydney University, and Dr Thomas Loy [1] of Queensland University. Both supported the interpretation of Dr Lawrence: that the material appeared to be skin tissue. However another scientist I engaged, Dr John Walker of Sydney University, had a different opinion. He said that the material looked like muscle tissue, not skin. So then I flew to Italy and sought the opinion of Professor Odoardo Linoli. He also thought it was muscle, perhaps even heart muscle, but couldn't be sure. He was making his assessment based on the photographs I was showing him. He wanted to examine the slides themselves under his own microscopes before coming to a definite identification.

I had a problem. Three scientists said it was skin. One said it was muscle. One said it may be heart. The same slides were being identified by different scientists as completely different tissue. I could have simply concluded that the majority opinion carried the day and that if three out of five said it was skin then it was skin. Except that's not how science works. It's not a democracy. It doesn't matter how many scientists concur if they are all wrong. Remember Galileo Galilei. His was a lone voice contradicting 97% of his scientific peers who were certain the earth was the centre of the solar system. No matter how many said it was so didn't make it so. I did what I could. I embarked on a dedicated program of self-education using the internet. For a whole year I gained access to university portals and undertook in depth studies of forensic pathology and cell biology tutorials. I was learning a lot but getting nowhere. Until one Saturday night.

On my knees in front of the tabernacle at Mass I prayed. "Jesus help me. The experts haven't been able to help me. I don't know where to go. You promised you would help. I need your help." The next morning I carried on working. I remembered a throw away comment Dr Walker had made about the statue sample, something about something weird looking like the nuclei of liver cells. So even though my problem had nothing to do with understanding the samples taken from the statue and everything to do with the samples from the Argentina host, I did a Google search: 'nuclei of liver cells + histology + images'.

Four photographs of microscopic images of human tissue came up on a University website. One showed the nuclei of liver cells. It was of no consequence. But the one above it hit me like a ton of bricks. It was an almost perfect copy of the photos taken from the microscopic slides of the Argentina communion host. The text accompanying it explained a lot. It read:

'This micrograph shows *irreversible cell injury in the heart* (acute necrosis in a myocardial infarct). The nuclei of the muscle cells are not staining, and there are many polymorphonuclear neutrophils infiltrating this tissue; *an acute inflammatory reaction'*.

I had keyed in words pertaining to the statue case and instead, by pure chance, had identified the tissue in the Argentina case.

My subsequent research revealed why. General microscopic depictions of heart muscle in pathology atlases show the tissue as it would be in *normal healthy* conditions. But the material on my slides showed little correspondence with these standard representations. This new information was alerting me to the fact that *heart tissue changes in appearance when traumatised.* One of the telltale signs of injury is inflammation, a proliferation and infiltration of white blood cells into the tissue. They turn up like warriors to fight the injury or the enemy.

The normally elongated smooth striated appearance of the heart muscle tissue changes. The tissue loses its striations, separates, becomes curly and eventually disintegrates. The material on the slides looked nothing like healthy cardiac tissue but exactly like textbook traumatised cardiac muscle replete with an abundance of white blood cells.

This new revelation changed the whole course of my investigation. If it hadn't happened, all my work, and all that it means for human history, may have stalled and faded into oblivion. What I discovered by accident, if there is such a thing, set a reference point for all scientific investigations of eucharistic miracles which followed. Buenos Aires was the first.

I knew exactly what I had to do next: Find the most qualified, most experienced forensic pathologist who specialised specifically in the human heart and ask him to analyse the samples I had taken from a piece of bread that had been in a test tube of water for three years in Buenos Aires. I found him in New York.

(1) Dr Thomas Loy, the only real-life scientist mentioned in Michael Crichton's famous book 'Jurassic Park', was himself an expert in identifying blood from ancient samples found on tools and weapons. He had analysed DNA in the ancient blood found on the tools and clothing of the 4000-year-old 'Iceman', Otzi, a frozen mummy found in the Alps. He wanted to identify what the 'Iceman' had hunted just before he died and was preserved in glacial ice. When I first met Loy he was working on a way to decipher ancient damaged DNA. Mike Willesee and I funded some of his research in this direction but before we could get useful results Tom Loy suddenly died . He was only 63.

CHAPTER 15

NO DOUBT ABOUT IT

Dr Frederick Zugibe had such an impressive list of qualifications in anatomy, electron microscopy, cardiology and forensic pathology that just by going through it I was absolutely certain he was the specialist I needed. He had been a forensic pathologist since 1969 and was so well respected that he even had a disease named after him. Zugibe spent most of his career as the Chief Medical Examiner of Rockland County, New York. He had written books on his work. He held a Bachelor of Science, Master of Science (Anatomy/Electron Microscopy), a PhD (Anatomy/ Histochemistry), and an MD degree. He was a Diplomate of the American Board of Pathology in Anatomic Pathology and Forensic Pathology, and a Diplomate of the American Board of Family Practice. Dr. Zugibe was an adjunct Associate Professor of Pathology at Columbia University College of Physicians and Surgeons and was a Fellow of the College of American Pathologists, a Fellow of the American Academy of the Forensic Sciences, Forensic Pathology Section, and a member of the National Association of Medical Examiners. He was formerly Director of Cardiovascular Research with the Veteran's Hospital in Pittsburgh, and was a Fellow of the American College of Cardiology, Fellow of the Council on Arteriosclerosis of the American Heart Association, Fellow of the New York Cardiological Society, and Member of the International Atherosclerosis Society.

After e-mail correspondence he agreed to meet me in person. I convinced Mike Willesee to come with me and so on 20 April 2004 we flew to New York with my samples and the sole purpose of seeking his opinion. Dr Zugibe lived in Ploughkeepsie, about 100 km up the Hudson River north of Manhattan. So we made our way to Grand Central, the landmark train station in New York City. Mike was elated to be on a train. Years of superstar celebrity status had entirely removed train travel from his life and he rediscovered a joy long forgotten.

When we finally knocked on Dr Zugibe's door he ushered the two of us into his office. He was very abrupt.

"I haven't got much time for you," he said, "so let's get right to it. What's all this about?"

On every occasion I engaged the services of scientists I never informed them of the source of the material before they delivered their verdict. This time would be no different. I explained that I was seeking his professional opinion on a case I was working on as a lawyer.

"I can't tell you any details. I'm sorry. But it would compromise my case. It's a very sensitive investigation."

He wasn't very happy with that answer.

"Well that's not good for me. No, that's no good for me. I've been a County Medical Examiner for more than 30 years. The police call me. I go to the scene of a crime, the scene of an accident. They say, 'It's a white male, six foot, 200 pounds, suspected gunshot wounds.' So I know a lot before I get there. And you're telling me nothing?"

We did not bow to his demands.

He retorted testily, "I always have a history before I start."

With his permission I then set up my camera, and started recording while Mike continued to talk with him. From these video tapes I can now quote the conversation that took place between the three of us verbatim and can visually show what Dr Zugibe was looking at when he was speaking.

"It's important you have a totally unbiased view of this," said Mike. "But we'll tell you everything after you've had a look at it." It wasn't what he wanted to hear but he proceeded to take the samples I carefully unwrapped. He placed the first slide under his microscope and placed his eye on the eyepiece. It took no more than a few seconds before he spoke calmly and confidently, without taking his eyes off of what he was seeing.

"I am an expert on the heart. The heart is my business. This is flesh. This flesh is heart muscle tissue, myocardium, from the left ventricle wall not far from a valvular area. It is the muscle that gives the heart it's beat and the body it's life. This heart muscle is inflamed. It has lost its striations and is infiltrated with white blood cells. White blood cells are not normally found in heart tissue. These cells are produced by the body and they escape from blood and infiltrate the tissue to address trauma or injury."

The academic in him described what he was seeing and also contextualised it explaining what was inferred and the implications.

"The heart of the person from which this tissue has come has been injured and has suffered trauma. The blood supply to this heart has been compromised resulting in necrosis, destruction of part of the heart muscle. This heart has suffered."

While I was careful to keep filming I remember distinctly how these words made me catch my breath. Dr Zugibe continued.

"This is not unlike what I see in motor vehicle accident cases when CPR is applied to a person too harshly and the heart suffers injury. It is also the sort of suffering that I see in cases where someone has been severely beaten around the chest."

Mike and I looked at each other. Mike asked Zugibe,

"And the white blood cells. What do they indicate?"

"They indicate injury and inflammation. Well there are a lot of them. All intact. White blood cells can only exist if they are fed by a human body."

Dr Zugibe paused and finished his explanation. "This sample was alive at the moment it was collected."

It was my turn to pause. I couldn't quite digest what I was hearing. I cleared my throat and asked,

"And how long would the cells remain vital if they were in human tissue that had been placed in water?"

Dr Zugibe said, "White blood cells cannot exist outside the body as they are fed by it. White blood cells will normally dissolve within minutes to an hour from being taken from the body. They would no longer exist."

He wasn't done. There was more.

"Furthermore, it would be impossible for the white blood cells to be present in the sample if the sample had been kept in water. The morphology of the tissue here showed good fixation. I would only expect to find such a good state of preservation of the tissue if it had been placed in a preservative like formalin."

As I registered the enormity of what Dr Zugibe had just said I zoomed in to a tight close up of his face peering down the microscope. He hadn't looked up from it all the while he was speaking which gave Mike and I free rein to communicate in silent facial expressions about what we were hearing. When I was ready I signalled to Mike.

"What would you say," he asked the professor, "if I were to tell you that the source of this sample had been placed in ordinary tap water for a month, and then stored for three years in distilled water before a piece was taken from it, and fixed on the slide you are now looking at?"

Zugibe looked up. "Absolutely unbelievable. No explanation can be given by science." He shook his head.

"And what would you say," continued Mike, " if I told you the source of this specimen was a piece of wheaten bread, a communion host?"

It was Professor Zugibe's turn to be shocked. He seemed to be scrambling for a logical explanation. But there wasn't one in his vast deposit of science and experience. He regained his professional composure and delivered a carefully spoken response for the camera.

"How or why a communion host could change its character and become living human flesh and blood is outside the ability of science to answer."

The snapshot of the heart on the slide was discernible to Dr Zugibe as a particular histological myocardial condition. It revealed evidence of cardiac tissue degeneration caused by trauma. From the appearance of the muscle structure, the particular type of white blood cells and the extent to which they were present, Dr Zugibe drew the conclusion that the trauma had occurred two to three days before.

What was recognisable to him and to anyone who looks at the slide, was the complex process of the human immune response at a defined point in a defined biological process. In this case it was severe trauma. The myocardial condition was not caused by the death of a person. Cells on the slide were not moving, and most definitely not jumping around, as some have erroneously stated. The slide showed no recognizable biological markers of death. "I am looking at a snapshot of a living person," he said. "There is no evidence of death."

In his written report to me he wrote, 'The slides contain cardiac tissue that displays degenerative changes of the myocardial tissue with loss of striations of the muscle fibres, nuclear pyknosis, aggregates of mixed inflammatory cells consisting of chronic inflammatory cells (macrophages) which are predominant and smaller numbers of acute inflammatory cells (white blood cells primarily polymorphonuclear leukocytes) which are admixed. The directionality of the myocardial fibres indicates that the site of these changes is relatively close to a valvular region in the ventricular area of the heart. These degenerative changes are consistent with a recent heart attack (myocardial infarction of a few days duration) due to an obstruction of a coronary artery that supplies nutrients and oxygen to an area of heart muscle. This obstruction maybe the result of atherosclerosis (process of fatty plaque build up) or a coronary thrombosis (obstruction of the coronary artery by a blood clot) or a severe blow to the chest over the heart.' [1]

I had to get my own thoughts straight. This heart expert is saying that the tissue was a snapshot of active human cell behaviour. It was evidence of life, not death. And yet what the professor is describing is tissue on a slide, prepared and fixed after having been in water for more than three years. This means that until the tissue was fixed as a slide, it must have remained as living tissue.

Why does Professor Zugibe say it was alive? Because white blood cells, particularly the clearly visible, intact, polymorphonuclear leukocytes, cannot survive outside a body for more than a few minutes before disintegrating. Indeed the entire life span of white blood cells is a short cycle mere days even when in a living body when they are actively addressing trauma in a tissue. And yet he was seeing them on a slide.

Dr Zugibe's assessment of the good state of preservation of the white blood cells and the muscle tissue was what might have been expected if they had been placed in a preservative like formalin. But the liquid substance in which the material sampled had been kept was analysed by Dr Thomas Loy of Queensland University on 30 August 2001.He reported, "No organic odour was detected that might be associated with histological fixing chemicals (commonly, alcohol, xylene, formaldehyde). The refractive index of the fluid was compared with water and known standards and determined to be essentially that of water."

Biology textbooks will tell you that all tissue, but particularly white blood cells, dissolves and disintegrates readily when placed in water. By osmosis water moves into the cells which then swell and explode, undergoing cell lysis, meaning cell destruction. They, or even remnants of what once were cells, would be very difficult to find under a microscope. Osmotic pressure on cells placed in distilled water is even higher than tap water and more detrimental to their longevity. Dr Zugibe confirmed this. "It would be impossible for the white blood cells to be present in the sample if the sample had been kept in water" and that, "if this tissue had been placed in water it would normally have broken down within one week."

Zugibe not only identified the tissue, he examined it forensically to provide an historical account, a picture of what had recently transpired in the heart from which the tissue is only an extract. He provided a timeframe. From what he saw he described what happened in the life of that 'person'. This 'person' experienced cardiac trauma two to three days earlier. Dr Zugibe could date the event that compromised the blood supply to the heart by the presentation of the tissue itself and by how far the process of immune response had progressed.

The issue of it showing trauma which occurred three days prior to the taking of the sample left me puzzled. But a theologian, Monsignor Hendryk Micek, wasn't puzzled by the three day issue at all. He explained that it made sense to him because after the consecration of the bread during the Holy Sacrifice of the Mass, the *resurrected* body and blood is made present and shared in communion. The state of the heart tissue reflected the trauma of the crucifixion experienced by a real historical man called Jesus of Nazareth three days prior to his resurrection. This was even more evidence for his faith in the Eucharist.

The five wounds which remained in his glorious body were not only the externally visible ones, the holes in his hands and feet and the wound in his side, but now also, the internal wounded physical heart.

Not only was the person 'alive at the time the sample was taken', but this person had a particular medical history because this particular heart tissue revealed, 'degenerative changes consistent with a recent myocardial infarction of a few days duration… due to an obstruction of a coronary artery that supplies nutrients and oxygen to an area of heart muscle. This obstruction may be the result of atherosclerosis or a coronary thrombosis or," said Dr Zugibe, *"a severe blow to the chest over the heart."*

I had a million questions. Most were a lot more puzzling and complicated than simply asking how someone could have had access to a locked tabernacle.

How could anyone have manipulated a piece of bread so that it transformed into what appears to be human tissue?

Had someone, say a surgeon or a murderer, sliced off a small piece of heart tissue from a living person and then dropped it into the bowl of water with the host, how could he then have engineered it to be embedded in the host so that it was of one substance with the wafer?

How could he have engineered its continuing transformation whilst securely locked away?

How could he have altered the natural rates of decomposition for cardiac tissue?

How could he have defied nature and kept the white blood cells alive for longer than a few minutes?

How could he have defied the physical laws of entropy and decomposition and kept the white cells alive for three years in water, cells which under normal circumstances would not last for more than a few minutes?

How could he have fiddled with the cell nucleus so that forensic laboratories identified it as clearly containing human DNA yet were baffled as to why they couldn't get the 'victim's' DNA profile? DNA, yes... DNA profile, no... Very strange?

How would he have maintained an ongoing process of tissue activity (inflammation) in a disembodied piece of heart whilst transporting it from a theoretical corpse/murder victim to the glass bowl in the tabernacle?

How could he have had that process of life continue after placing the tissue in water?

And since no human could survive having a piece of his heart cut out I just had to ask: Where is the rest of the heart?

Where is the body?

It was clear to me that I was involved in something I knew virtually nothing about. But I was set on learning whatever there was to know. At that time computers and the internet were still in their infancy. No one, neither in nor outside of my local Church, had ever heard of the strange phenomena I had seen. I didn't even know at that stage that communion hosts suddenly transforming had a name. And, as I found out, a history. So I started to read a lot and eventually spoke to some highly educated people about what were called, "Eucharistic miracles". How had people discussed these cases at the time? Had their conclusions been rational? How, if at all, had scientists been involved? And if there were documented historical Eucharistic transformations why didn't the Church talk about them?

---

(1) See footnotes to Chapter 8 of the book 'Unseen, New Evidence, the origin of life under the microscope' by Ron Tesoriero and Lee Han 2013 for the sources of material for this chapter including the supporting Scientific reports.

CHAPTER 16

LANCIANO

Lanciano is an Italian town named for the lance of one of its inhabitants, a Roman centurion who lived there 2000 years ago. It was his lance which pierced the heart of Jesus as he hung on the cross to ensure that he was dead. In 750 AD a transformation of a host, similar to what had happened in Buenos Aires, occurred in Lanciano, in full public view, during a Mass in the Church of St Legontian. The priest, ordained in the order of St Basil, held up the altar bread at the moment of consecration and said the prescribed prayer. He had sincere personal doubts about the central and abiding claim of the Catholic Church. He couldn't accept that the body and blood of Jesus become truly and substantially present, body, blood, soul and divinity, at the moment of consecration. But these thoughts he kept to himself. Outwardly he was simply going through the motions, doing and saying liturgically what priests had been doing for centuries and what he himself had done hundreds of times before. But on one particular day something unprecedented happened:

In full view of the congregation the bread in his hands changed into what appeared to be flesh and the wine in the chalice appeared to turn into blood. This 'blood' later coagulated into five irregular pellets, each a different size and shape. The transformed bread and wine were preserved and subsequently sealed in separate glass vials. Elaborately bejeweled reliquaries were commissioned to house these glass vials and for centuries they were displayed as treasured objects of veneration by pilgrims.

In 1971, more than a thousand years later, the custodians of these relics, the Conventual Friars Minor, requested a modern scientific examination of the contents of the reliquaries. The investigation was authorized to be undertaken by Professor Odoardo Linoli, Professor in Anatomy and Pathological Histology and in Chemistry and Clinical Microscopy at the Arezzo Hospital in Italy. He in turn requested that the slides of the samples be prepared by Professor of Anatomy, Ruggero Bertelli, of the University of Siena.

What they discovered and published was that the section taken from the 'host' was found to be human tissue. It was identified as tissue from the muscle of the heart. What was believed to have been wine was identified as human blood, Type AB. Both the blood and the heart tissue had remained, for twelve centuries in their natural state, inexplicably preserved, since no preservative was detected.

In 750 AD, Lanciano Italy at the time of consecration during Mass, bread appeared to turn to flesh and the wine to blood. For 12 centuries the substance remained in a sealed reliquary until 1971 when it was scientifically examined.

On 2 March 2001, Prof Odoardo Linoli explained his tests results to Mike Willesee and myself. He said **"This is Human heart tissue."**

Professor Linoli's research and analysis was first published in 1971 in Volume 7 of Quaderni Sclavo di Diagnostica Clinica e di Laboratori. It was subsequently published in Osservatore Romano in 1982 and then summarized and updated in the Fr. Nicola Nasuti's 1991 publication, 'The Eucharistic Miracle of Lanciano'. I was intrigued enough to book a flight to Rome and personally interview Professor Linoli. I filmed the interview on 7 March 2001 Professor Linoli confirmed all the results of his studies detailed in his earlier report.

"Has your work been verified?" I asked.

"The documented results, including photographs and a detailed methodology, were published and have been open to public scientific scrutiny, and yet in thirty years no one… Absolutely no-one has disputed or contested the conclusions, analysis or the diagnostic methods used by Professor Bertelli and myself."

He reaffirmed unequivocally that the substances examined were human flesh and blood. The blood was of the rarest blood type, AB, a type with geographic affiliation to the Mediterranean. Proteins found in the blood were present in same proportions (as percentages) as those found in the sero-proteic make-up of fresh normal blood. Typical minerals found in human blood were present: chlorides, phosphorous, magnesium, potassium and sodium.

The flesh consisted of the muscular tissue of the heart. Coincidentally, the same part of the heart described in the Buenos Aires case.

This concurs with the findings of Professor Bertelli who wrote in his report of 26 February 1971 that, 'The examination of the glass slides has suggested to me that the microscopic sections, thereon contained, are constituted by a tissue that is mesodermic in origin, and more precisely by a striated muscle tissue. The course of the muscular bundles, which intersect on the different layers and in diverse directions, the presence of anastomatic areas between the fibres would lead to the identification of a myocardial muscular tissue.'

Present also, in transverse section, were the synctoid structure of the myocardium, the endocardium, a branch of the vagal nerve replete with its thin perinium and recognizable fascicular structure, both arteries and veins within the myocardial tissue and also the left ventricle. That is to say that what was visible was a comprehensively representative slice of the essential structure of a complete heart.

Professor Linoli was fascinating to talk to. He explained his work in great detail, particularly the experiment comparing the capillary action of the liquefied blood from the reliquary with fresh blood. He explained the law of histology used in his assessment.

"If an unknown substance," he said, "behaves the same way as a known one then the law of histology states it is the same. When liquefied, the substance in the reliquary matched exactly the capillary properties of freshly shed blood. It corresponded with human blood taken from a man's body that very same day."

Particular emphasis was made by Professor Linoli of the remarkable state of the flesh and blood.

"Considering its age and the natural processes of decomposition it was remarkable and scientifically inexplicable. No traces were found in the organic tissue of any substance intended to preserve it. Even so, in spite of thirteen centuries of exposure to the action of biological and atmospheric agents, they had remained pristinely preserved."

Aware of the fact that the full impact of his findings may not be fully appreciated by those not trained in the sciences of anatomy and forensic pathology Professor Linoli commented that although it was outside of his objective scientific task strictly speaking, there was something else to consider.

"I couldn't help but note that only a highly skilled hand in anatomic dissection could have obtained such an 'even and continuous' slice of heart tissue. This was all the more intriguing considering that the very first anatomical dissections in the general historical literature occurred only after the 1300's, almost six hundred years after the Lanciano event."

For Linoli this observation dismissed any possible charges of a fraud having been perpetrated in 750AD.

Like the Argentina case, it was as if science alone was demolishing any hypothesis of fraud.

CHAPTER 17

THEN IT HAPPENED AGAIN

Sokołka (pronounced Sokooka) is a very small town only just inside Poland's eastern border with Belarus. It lies on the railway line that connects Warsaw with Moscow. Sokołka is the very last stop before leaving the European Union. The picturesque little town has been tossed between the Soviet Union, Nazi Germany, Poland and Belarus as one of the spoils of war for much of its history. It has experienced the very depths of human misery particular to the twentieth century. This frontier town has borne the brunt of multiple wars, of brutal foreign invasions, by the Germans and then by the Soviet Union, of tyrannical dictatorships, and of the devastation wrought by Nazism and Marxism.

Until the outbreak of the Second World War the region, which includes Bialystok, the biggest town, held a demographic record: It was inhabited by the largest concentration of Jews in the entire world. Approximately 75 percent of the population was Jewish.

What proved therefore even more deadly than the military invasion of Poland by Germany was the genocidal anti-Semitic ideology of the Nazi occupiers. They proceeded to systematically murder the Jewish population. Records from Treblinka, only one of many notorious death camps, show that in it alone 109 000 Jews from the Bialystok region alone were exterminated; 8000 were deported to Treblinka from Sokołka in a single day.

After the Germans the Russians invaded. And stayed. The Russians remained in occupation in Poland and Sokołka found itself on the wrong side of the Soviet Union's Iron Curtain. Its majority population of Jews had been virtually annihilated and their property destroyed or confiscated. All that remained of the little town were a few historic buildings: the eighteenth century mosque, the unique wooden Jewish shul, the Orthodox Church with its revered icons and the 1848 Catholic Church of St Anthony of Padova. By 2008 the Soviet Union had collapsed, but the old places of worship were all still there.

It was in the Church of St Anthony that a young assistant priest, Father Jacek Ingielewicz, accidentally dropped a consecrated communion host during Mass on the 12 October 2008. He picked it up and noticed that it was slightly soiled. He did what he was trained to do. He placed it in a special liturgical vessel called a vasculum, added some water and locked it in the tabernacle. Directly after the Mass, on the same day the parish priest, Father Stanislaw Gniedziejko, noticed the vasculum in the tabernacle containing the host. He instructed Sister Julia Dubolka, the sacristan (someone charged with the care of the sacred vessels, vestments and candles), to empty the contents of the vasculum into a glass bowl and to place the bowl in the safe in the sacristy behind the main altar of the church. She did as she was asked and then she locked the safe.

Seven days later, on 19 October 2008, Father Stanislaw asked Sister Julia if the host had dissolved. Sister Julia then opened the safe and discovered that the host exhibited an unusual change. A red 'stain', she said, was visible and gave the impression of being blood. She went and told Father Stanislaw who came to see it for himself.

"I was shocked and I didn't know what to think of this. My hands were shaking when I locked the safe. I could hardly speak."

Father Stanislaw then informed his superior, Archbishop Edward Ozorowski, who made the journey to Sokołka himself a few days later with his Chancellor, Father Andrew Kakareko. On 29 October 2008, on instruction from the Archbishop, Father Stanislaw removed the stained host from the water and laid it on a linen cloth called a corporal. The cloth was then placed in the tabernacle of the chapel in the rectory and locked.

The Archbishop appointed a special commission to investigate the unusual transformation with the specific intention of ascertaining whether there had been any interference with the host. On 5 January 2009, two respected pathomorphologists from the Medical

University of Bialystok were formally requested to conduct an independent analysis of the material. Professor Sobaniec-Lotowska and Professor Sulkowski hold chairs in different departments within the university, one in the Department of Medical Pathomorphology and the other in General Pathomorphology. They have been working as specialists in histopathology diagnostics for over thirty years. Their individual scientific achievements are prodigious and recognized beyond the borders of Poland. Both have provided the scientific world with a formidable body of valuable peer reviewed research.[1]

Professor Sobianic-Lotowska excised a section as a sample, about a square centimetre in size, in the presence of the Chancellor of the Metropolitan Curia of Białystok. She reported that she "securely took hold of this small piece of matter; it was brittle, of brownish colour with some remains of the communion host attached. I stress that at that time it was unknown to us what kind of material we would be dealing with."

But by the time the suite of tests had been concluded the final report compiled by the two independent pathomorphologists was unambiguous. "What is most important, in our opinion," said one of them, "is that the material analysed consists entirely of cardiac tissue."

When interviewed Professor Sobaniec-Lotowska was asked, "What observations made you certain that what you found in the tested sample was actually heart-muscle tissue?"

She replied that, "Even in such a small quantity of the communion host, we could observe so many typical bio-morphological indicators of heart-muscle tissue. One of those indicators is segmentation or damage to myocardial fibres at the site of intercalated discs as well as the phenomenon of fragmentation. Such damage is visible as tiny ruptures." For the professor, "Another important proof that the analysed material might have been a part of a human heart was the central position of the nuclei in the cells of the tested material, a distinctive trait of this type of muscle. In the structure of some of the fibres we could find impressions, which might be referred to as spasms in the lymph areas. Also, under the electron microscope we observed the outlines of intercalated discs and the outlines of bundles of delicate microfibres."

She described the sample as being cardiac muscle, "just before death. It is in agony, a moribund condition, caused by great stress. This is proved," she said, "by the presentation of a very strong phenomenon of 'segmentation' or damage to myocardial fibres at the site of the intercalated discs, which does not occur after death. Such changes can be observed only in living fibres and they show evidence of rapid spasms of the heart muscle in the period just before death."

Both scientists highlighted other particular observations which were, in their experience, extraordinary. Professor Stanisław Sulkowski pointed out that, "Under normal circumstances when we place the host in water it normally dissolves in a short period of time. In this case however a part of the host, for unknown reasons, did not dissolve."

Professor Sobianic-Lotowska asked that attention be paid to the surprising absence of autolysis, the process whereby a cell is destroyed by the action of its own enzymes, enzymes which are released due to the cessation of active processes in the cell. Autolysis usually occurs in injured or dying tissue. The fact of its absence in the sample astonished her.

"Additionally, please note yet another incredible phenomenon. For a long time, the host remained submerged in water and then even remained longer on the corporal and therefore the tissue, which appears on the host, should have undergone the process of autolysis but we did not observe any such changes during our tests. It is my opinion that according to the current state of knowledge in biology we cannot explain this phenomenon scientifically."

Another phenomenon, described as interpenetration or inter-absorption merited particular attention by both scientists. "What is even more difficult to comprehend," says Sulkowski, "is that the tissue, which appeared in the host, was closely bound to it, to the host that is, penetrating the base on which it appeared. Please believe me that even if someone intended to tamper with the sample, it would be impossible to bind the two pieces of matter in such an indissoluble way." Professor Sobaniec-Lotowska also found this aspect of the results remarkable. "This extraordinary phenomenon of inter-absorption of the heart muscle tissue with the communion host, observed under the microscope, and also via transmission by an electron microscope, proves to me that there could not be any human interference with the sample."[2]

The transmission electron microscope directs an electron beam through a section of the specimen and provides a two-dimensional cross-section of the specimen. Scientists can see a transverse view through the material as if cutting a slice of cake and seeing all the layers. What they saw in this case was an inextricable interweaving of human tissue with the material components of the host.

In August 2010 Mike Willesee and I travelled to Sokolka to interview the people involved in this case. Professor Sobaniac-Lotowska summarized the results of the investigation to me in an interview on 13 August 2010 in Sokolka. She verified her explanations by referring to pathology images.

The host had transformed, she confirmed, into "heart tissue showing structures common in spasm which precedes death. The cardiac impact, she reiterated, had been recent. The heart was alive, just before death. The sample analysed was not from a dead person. The person was alive. There was 1 square cm of heart. A fragment of muscle. If one had to remove it from a person, he would die.

Pointing to a photograph of cardiac tissue she repeated the fact that the fragment had been in water for weeks and that, "after only a single week this would no longer be able to be seen."

(1) See listed research on Pub Med database...PubMed database http:///www.ncbi.nlm.nih.gov/pubmed, U.S National Library of Medicine, National Institutes of Health.

(2) Extracts from the pathological findings of Professor Sobaniec-Lotowska and Professor Sulkowski were published in Nasz Dziennik in Poland 12 December 2009, no 291 (3692) pp 1, 9 and contained within the extensive interview by Adam Bialous. The publication of the report was translated into English from Polish by Janusz Tydda and then submitted by Ron Tesoriero to Professors Sobaniec-Lotowska and Sulkowski for confirmation of its accuracy and correct translation of all technical content. The confirmation and corrections were given to Ron Tesoriero by the Professors on 28 June 2011. The quotations from the scientists are from this interview and from an interview filmed by Ron Tesoriero and Mike Willesee on 13 August 2010

At the Church of St Anthony, Sokolka, at Mass a Communion host dropped to the floor. It was placed in a bowl of water and locked in a safe behind the main altar. (12.10.2008)

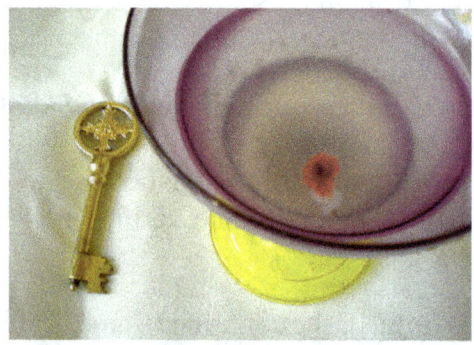

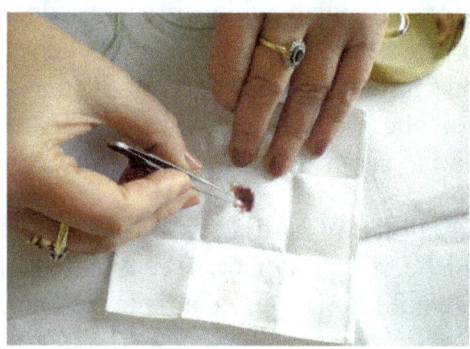

When the safe was opened, the host had changed exhibiting a blood-like substance. (19.10.2008)

Prof Sobaniac Lotowska and Prof Sulkowski explained to Mike Willesee and myself when discussing the test results. **"This is heart muscle tissue from a person who has suffered trauma."** *(13.08.2010)*

Archbishop Edward Ozorowski accepted the scientific results and proclaimed it to be a supernatural event at a special formal ceremony. (02.10.2011)

CHAPTER 18

SHE SAW WHAT SHE WANTED TO SEE

The Polish Rationalist Society was quick to comment on the remarkable events in Sokołka. If, they declared, living heart tissue had been found then a murder had been committed somewhere and a search for the body should begin immediately. They took the matter to the Public Prosecutor of Sokołka. He responded with a declaration. There was no evidence of any murder in the town.

Professor Lech Czyczewski wore two hats; he was the head of the department of Medical Pathomorphology in which Professor Sobaniac-Lotowska worked, basically her boss, and also the media spokesman for the university. He disputed the scientific report on the Sokolka host and the media quoted him accusing his two fellow scientists of involvement in 'illegal procedures', that is, of conducting tests and procedures on their own initiative and on their own account, and of compromising their scientific integrity with their religious beliefs.

When confronted by the media regarding these accusations Professor Sobaniec-Lotowska provided two documents in her defence. The first was the original formal written request by the Chancellor of the Curia of Bialystok to undertake an investigation of the inexplicable phenomenon of the 'stained' host. Clearly she had not acted on her own initiative. The second was a letter from Dr Janusz Mieczkowski of Szczecin, an eminent Polish academic. She read an excerpt:

"A formal request or lack of it, under no circumstances should decide the validity of the pictures of the microscopic slides taken. The university spokesman does not question the actual scientific analysis and the diagnosis yet at the same time adds a commentary which 'accuses both scientists of an emotional relationship to their religious faith.' He does not specify what facts made him formulate such a statement. He expresses the opinion that 'both scientists made a basic error of mixing theology with biology, which is not possible.' It is difficult to agree with such a position. If indeed the co-existence of biological science with theology was not possible we would not have any religious persons among educated people. The finding of heart muscle tissue in the communion host is evidence of a miracle and there would have to be a good evidence to prove otherwise."

Professor Sulkowski, the other pathologist in the case who also came under a barrage of criticism, reflected on the larger issue at stake questioning the role of the scientist in society? He asked, "It is also my opinion that if there is an important social issue, which requires the involvement of a scientist, if there is need for his knowledge, it is not only his right but his duty to participate. I look at it as kind of a service for society, which funds our scientific activities. Society has the right to all the possible information arising from all branches of human knowledge. As no medical practitioner can refuse to help the sick, so it is that we also have the duty to research every scientific problem according to the guidelines outlined by the Committee of Academic Ethics of the Polish Academy of Sciences."

The results of the two scientists' findings however continued to be undermined, this time by the press. The headline of an article published in 'Super Express' read, 'Not a miracle. Pure Biology.' The journalist who filed the story quoted Dr Pawel Grzesiowski, a biologist employed by the National Institute of Pharmaceutical Drugs, disputing the finding of heart tissue: 'What was found in the church was the serratia marcescens bacterium. The serratia marcescens bacterium is commonly found in bathrooms and feeds off of starch and sugary substances. It manifests itself as a pink, slimy substance. The host most likely got damp in the tabernacle where it was stored after the Mass.'

The article also included a statement by yet another scientist Professor Lech Chyczewski, head of the blood unit at the Bialystok hospital. Of Professor Sobaniec-Lotowaska he said, 'The professor saw what she wanted to see. She is very religious.'

Both the investigating professors who analysed the sample remained adamant. "We are certain about the outcomes of the tests because we conducted them meticulously. If it weren't for the Curia's requests, we would have initiated legal proceedings against some of the media outlets, which have published obvious lies. That the gentleman (referring to Dr Grzesiowski and his 'pink slime' diagnosis) stated an opinion on an object, not seen by him even once, not to mention that he never lab-tested it, is unethical. I think that his attempt to explain these phenomena was pseudo-scientific and it comes from his total lack of knowledge of what was actually involved in the tests." Professor Sulkowski added that, "Dr Grzesiowski's conviction that such bacteria existed in the material he had never seen in his life is strange to say the least."

I had known the same kind of frustration. Cynics would routinely ridicule the results and yet not once, in the twenty years I had been investigating these phenomena, had I been asked by these same critics if they could examine the actual material or the actual scientific reports or the actual video evidence. The calibre of logical investigation and curiosity was so diminished that it was acceptable to declare that these things can't happen and therefore they didn't happen.

I remembered my interaction with Dr Susanne Hummel, a heavyweight in the field of DNA research and author of a definitive textbook on DNA typing.

I requested, via email, that she examine my sampled material from the Buenos Aires host. My standard line was that I was a lawyer working on a case and I needed to engage her professional expertise. I couldn't disclose the source without compromising the results.

When Mike and I met her on the 15 December 2005 at the University of Göttingen in Germany she insisted we tell her the source otherwise she wouldn't take on the case. So Mike told her the history of the sample. Her associate said words to the effect that there would be no point in doing the tests because they knew the results without the need for test. He said that it was most likely some form of fungus or bacteria that had developed on the host. There are some bacteria forms that are red.

Then Dr Susanne Hummel said something which entirely stunned me. I was rudely awoken from my naïveté that science was professionally unbiased and ideologically neutral. She said that they would not take on the case because if the testing confirmed what was being claimed, namely that a communion host had become flesh or blood, then this would embarrass the university and force it to close down some of its programs and activities were founded on atheism, (the University of Göttingen is internationally renowned for neo-Darwinism.)

The Polish professor defending the scientific integrity of her results concluded plainly that, "No bacteria known to science can actually produce matter with the characteristics of heart muscle tissue and this is what we have found in the host. If someone doesn't want to believe it, even if he sees it, he will close his eyes."

CHAPTER 19

THE MIRACLE OF MEXICO

When claims that the same spontaneously transforming communion host phenomenon occurred in yet another country, and on another continent, I was naturally intrigued. This time it was in a very small rural town in Mexico called Tixtla. Tixtla lies about 300 kilometres south of Mexico City, in a part of Mexico gripped by the violence of drug cartels.

In October 2006 a retreat for six hundred people was being held in the parish of St Martin of Tours. Holy Mass was celebrated, the host was consecrated and during communion a religious sister assisted with distributing communion. She reached into the bowl and picked one up. It looked as if it was 'bleeding'. Taken aback, she immediately replaced it in the bowl with the remaining communion hosts. Filled with emotion, her eyes brimming with tears, she turned to the presiding priest, Father Raymondo Reyna Esteban. He came over to have a look. He then lifted it up and, in full view of hundreds of witnesses, showed the host oozing a red substance.

The local Bishop of Chilpancingo-Chilapa, Bishop Alejo Zavala Castro, instituted a formal investigation three years later. At a symposium, broadcast live to 90 countries by Mexican television, the scientists involved in the Tixtla case presented their findings. Prominent amongst them was Dr Eduardo Sanchez Lazo, a noted forensic investigator, surgeon, an expert in Medical Law, and Professor of Legal Medicine from the Faculty of Medicine at UNAM, the National Autonomous University of Mexico. At the close of his impressive presentation he formally submitted his report to the Bishop. The Bishop officially pronounced on 12 October 2013, "I recognise the supernatural character of the series of events relating to the bleeding host of Tixtla… I declare the case a 'Divine Sign."

Exactly what led him to this conclusion was explained to me in person by Dr Sanchez himself when Mike Willesee and I travelled to Tixtla to meet him. Getting there was difficult. Security for the journey was provided by a truckfull of men with rifles travelling alongside us all the way to the doors of the Church. I spoke to the priest involved and examined the original host, carefully protected behind bulletproof glass.

The tissue was identified as heart muscle tissue. The red stains on the host were human blood. Blood type AB. Both white and red blood cells were evident. Upon close microscopic examination his team concluded that the blood did not originate from the outside of the host. Instead it originated from a central point within the host itself. It was not unlike what happens, he explained, when there's a wound. The skin is injured, a blood vessel is pierced and the blood flows from a single source, the vessel, and then spreads more widely as it exits the wound onto the surface of the skin. Dr Sanchez confirmed that this observation was one of the most significant in his determination that the blood was not placed on the surface of the host by someone. It had an interior cause, not an exterior one.

And then, finally, he reported the perplexing DNA results, which matched the other equally enigmatic cases I had tested. Human DNA was present. It was present in more than adequate quality and quantity to generate a human genetic profile. And yet they were unable to generate a genetic profile. The actual physical material was present. The scientific means to test it were present. Thousands, if not millions, of similar standardised DNA tests had been performed across the world over many years, and yet none had resisted providing a human genetic profile. Why?

In October 2006, in the parish of St Martin of Tours, Tixtla, Mexico during a mass while communion was being distributed, one of the hosts was seen oozing a red substance.

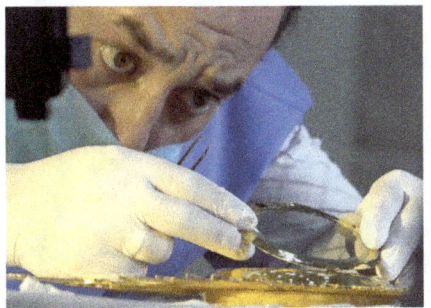 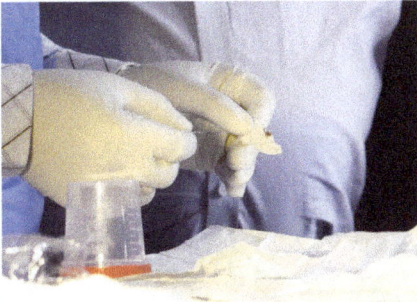

On 15 September 2015, Dr Eduardo Sanchez Lazo, forensic investigator and Professor of Legal Medicine, involved in the investigation called for by the local bishop, told me that he had reported a finding that **the substance was human blood and heart muscle tissue.**

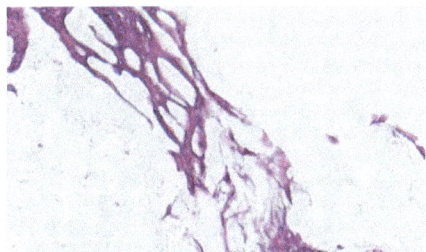 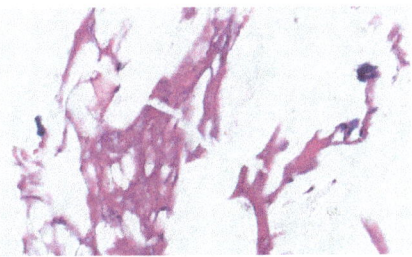

Microscopic sections showing evidence of heart muscle tissue.

CHAPTER 20

LEGNICA

In 2013 it happened again. Once again in Poland, but this time in a different town, Legnica. It is an ancient Silesian town situated in the middle of Poland. On Christmas day in 2013, during Holy Mass in the Church of St Hyacinth, an assistant priest began distributing the consecrated host to the altar servers behind the altar. One of them fell to the ground. Once again, following protocol, the host was picked up, place in a container of water and secured in the tabernacle.

In an interview I filmed, Father Andrzej Ziombra, the parish priest said, "From time to time while saying mass in the next days I had noticed the host in the bowl, but saw nothing unusual. But on the 4 January I saw the host in the bowl of water had become reddish. I then contacted the Bishop, Stefan Cichy, and told him what had happened. He said to me to wait another two weeks to see if anything further happened. After two weeks I noticed that the part of the host that did not have the reddish colour had dissolved, but the part with the red on it had not dissolved."

I asked Father Andrzej, "What makes you think it was a genuine happening and not some fraud?"

"The transformation had occurred in a locked tabernacle and only four priests had a key. I know the three priests well. I don't think they would act fraudulently."

A short while later, in February, the Bishop of Legnica appointed a commission to pursue a scientific examination of the transformed consecrated host. Dr Barbara Engel, a cardiologist, was one of the scientists involved. I interviewed her in Legnica too. She confirmed that the substance was examined by the Department of Forensic Medicine at the Pomeranian Medical University in Szczecin. Once examined it looked like fragmented cardiac tissue. This observation, and not the presence of white blood cells, led the scientists to conclude that the heart had suffered grave trauma. She used the word "agony".

Nuclear DNA testing was done by the Forensic Genetics section of the university but strangely no genetic profile was obtainable. Mitochondrial DNA tests did produce some results including information which suggested that the source was Middle Eastern.

We pored over the photos she had supplied so I could examine them in relation to the others I had seen. She remarked that every one of the scientists involved thought that what had happened was amazing. Each one was really touched by what they found. They had not believed something like this was possible.

I asked her, as a scientist, "What do you say when you are confronted with a situation where a piece of bread has changed to heart muscle?"

She very calmly replied, "Science can't explain this. To try to explain this we have to enter a very different territory."

The new bishop, Bishop Zdigniew Kiernikowski, delivered the report to the parishioners of Legnica on 17 April 2017. He said what was found in the samples was human tissue most resembling cross striated heart muscle in trauma. He pointedly used the word 'agony'. He concluded by saying, "I presented the whole matter to the CDF (the Congregation for the Doctrine of the Faith) in Rome. Today, *according to the recommendations of the Holy See*, I ordered the parish priest, Andrzej Ziombro, to prepare a suitable place for a display of the relic so that the faithful can give it proper veneration and adoration. I hope that this will serve to deepen the cult of the Eucharist and will be unmistakable impact on the lives of people approaching the relic. We read this marvellous sign as a particular expression of the kindness and love of God, who so descends to man."

On 25 December 2013, in the Church of St Hyacinth, Legnica Poland at communion time, a consecrated host fell to the floor. It was placed in a chalice containing water. After 10 days a blood - like substance emerged.

Dr Barbara Engel

Cardiologist Dr Barbara Engel of Legnica was appointed to head an investigation. On 23 July 2016 I interviewed her. She said, "The substance was examined by the Department of Forensic Medicine at the Pomeranian Medical University in Szczecin and was found to be **fragmented cardiac tissue. That heart had suffered severe trauma.**" She provided me with these microscopic images showing evidence of the findings.

Bishop Kiernikowski of Legnica ended his report of 17 April 2017 saying, "We read this marvellous sign as a particular expression of the kindness and the love of God, who so descends to man."

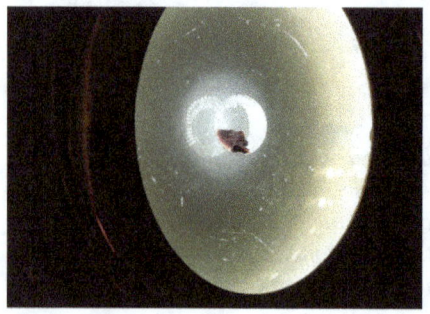 

Bishop Kiernikowski ordered that the relic be displayed so that the faithful may give proper veneration.

CHAPTER 21

MANKIND'S GREATEST INTELLECTUAL ACHIEVEMENT?

Being at the epicentre of the Argentina case meant travelling around the world for many years to continue scientifically examining the host samples, studying histopathology late into the night in order to interpret slides and results, exploring corollaries to the findings, and documenting the other cases, and making documentaries to share what I have found, has been the work of the best part of my life. It has been an all-consuming vocation. It has cost me. In money, yes, but even more in a commodity far more precious: time. I have paid in time lost, time not spent with my wife and family. I have relegated career and commercial interests to the side. My law practice was reduced to one small corner of my available time. What might be construed as sacrifices made by me were in fact minuscule obstacles.

The real disappointment for me has been how many friends, and even some members of my family, have dismissed the work without even wanting to look at the evidence. Imagine yourself as a lawyer who goes before a judge in a court case and tells the judge the general nature of your case. The judge declares, "I don't want to hear about it. I dismiss your case. I rule against you. I have already made up my mind. I don't want to hear any evidence you have to produce." Friends and family who are also lawyers should empathise with my sense of frustration. My disappointment is amplified by having prepared for so long and so meticulously to collate the evidence.

So single-mindedly wrapped up in this work have I been that I hadn't really considered exactly the full implications of the many scientific reports I had to consider. That is until 2009, when every time I turned on the TV there was yet another documentary celebrating a nineteenth century Englishman, a naturalist, who "forever changed the world". It was two hundred years since Charles Darwin was born and we were inundated with celebrating his life, his theory, and his famous little book, *The Origin of Species by Means of Natural Selection.*

Darwin and his Theory of Evolution were familiar to me but until then I hadn't read the book or closely examined what it proposed. When I did, one line he wrote took my breath away.

"If it could be demonstrated that any complex organ existed, which could not possibly have been formed by numerous, successive, slight modifications, my theory (that is the theory of evolution by natural selection) *would absolutely break down."*

Why did this line make such an impact on me?

For the last twenty years I had been investigating the spontaneous coming into existence, without numerous successive slight modifications, human heart tissue. The heart, by anyone's estimation, is a complex organ if ever there was one. There was no natural selection happening. The bread wasn't becoming a plant cell, then a worm cell, then a bird heart cell, then a chimpanzee heart cell, then finally a human heart cell. There was no succession. There was no parentage. There was no ancestry. There was bread and then there was human heart.

I thought it strange that no-one had ever before considered the logical implications of these host-to heart cases on Darwinian evolutionary theory. Had these tiny little samples and the scientific analyses done on them, caused Darwin's Theory of Evolution, in Darwin's own words, and according to Darwin's own test, to absolutely break down? Were these cases the beginning of the end of the Theory of Evolution?

No-one else was doing the work of finding answers to these questions and so I took them on myself. I knew it was important for me to first present the evidence and then show where that evidence lead. So I put on my lawyer hat and proceeded to analyse the data I had collected as if it were evidence in a legal case.

To be fair I had to understand exactly what Darwin's theory was proposing, what contemporary biologists were teaching, and what evidence was used in arriving at their conclusions.

Darwin claimed, and so did every single one of the thirteen university biology textbooks I examined, that life arose spontaneously on earth by means of an accidental fortuitous collision of non-living molecules. They called this "chemical evolution". All life, they unanimously claimed, had its beginning in a primitive ancestral cell, or cells, that arose spontaneously nearly four billion years ago. So in a nutshell, all life came from other life except for the very first instance of life. That very first cell, the ancestral cell, was the exception to an otherwise ironclad law of biology, originally proclaimed by Louis Pasteur, that all life comes from other life.

Way back in time when earth was cosmic dust, water, heat, and light turbulently directed by the laws of physics, by a process of 'chemical evolution' the very first living thing, supposedly a single cell, came into existence from non-living chemicals. And since then from this single cell each form of life evolved. Exactly how or why or where or when this happened is not addressed by Darwin or the textbooks. Neither is there an intellectually satisfying definition of exactly what "chemical evolution" is. What is claimed however is that it is true because it has to be true, or is most likely true, or is probably true. Take your pick.

The process of evolution through genetic mutation was, Darwin claimed, radically creative. It was so radically creative that it was the mechanism whereby entirely new species came into being. Plants evolved into animals, reptiles into birds and primates evolved into human beings. And it all started from the accidental existence of a single cell way back billions of years ago in what Darwin described as "a warm little pond with all sorts of ammonia and phosphoric salts, light, heat, electricity etcetera present." All that was needed after this first cell origin event was the material stuff already present, the laws of physics already present, and millions and millions of years, not already present, to account for all life, plant and animal, on this planet. This theory is applauded as "mankind's greatest intellectual achievement." Professor Richard Dawkins, a vocal neo-Darwinian biologist claims that,

'Evolution is a fact. Beyond reasonable doubt, beyond serious doubt, beyond sane, informed, intelligent doubt, beyond doubt evolution is a fact. The evidence for evolution is at least as strong as the evidence for the Holocaust, even allowing for eye witnesses to the Holocaust. It is the plain truth that we are cousins of chimpanzees, somewhat more distant cousins of monkeys, more distant cousins still of aardvarks and manatees, yet more distant cousins of bananas and turnips… continue the list as long as desired. That didn't have to be true. It is not self-evidently, tautologically, obviously true, and there was a time when most people, even educated people, thought it wasn't. It didn't have to be true, but it is. We know this because a rising flood of evidence supports it. Evolution is a fact.'

CHAPTER 22

THE ORIGIN OF LIFE

Darwin didn't simply add to our knowledge of the natural world, he provided biology with its fundamental modern philosophy: scientific materialism, the belief that the universe and what it contains, including life, is purely material, a product of natural law and chance. More significantly he provided atheists with the ammunition to kill off any 'spiritual' concepts of life and its origin. Biology graduated from being the science of the living world to being the authority underpinning secular society's principal ideology: that there is no such thing as God; that life was not created, that plants and animals were not created; that human beings are not created. Life, evolution reminds us, has no objective scientific meaning and no purpose. It just is.

Everything which lives on earth, including man, 'is the result of a purposeless and natural process that did not have him, or it, in mind.'[(4)]. Mind is just matter. Thoughts are just neurons firing, just chemical reactions. Even human consciousness itself is no more than a lucky roll of the chemical dice. We are, according to Christopher Hitchens, simply the most adaptable mammal of all with nothing more than a developed sense of 'pattern-making'. Our thoughts, our ability to reason, are as purposeless as the Creator they 'invented'.

But as science has become popularized for entertainment so too has the origin of life question spilled out of research laboratories and higher educational institutions into the life of the ordinary man. One night he watches a documentary in which scientists say that life was created in outer space and travelled to earth on a meteorite. The very next day another one makes the originalist 'warm little pond' claim that life started when the chemical ingredients of a pre-biotic soup assembled themselves by accident billions of years ago. On the third day a documentary assures him that life is the result of intelligent design. By the fourth day he's seriously considering whether life sprang into being in the cocktail of chemicals bubbling from boiling vents deep in the ocean. By Sunday he's back to the old story of creation: 'In the beginning was the Word, and the Word was with God, and the Word was God. The same was in the beginning with God. All things were made by him: and without him was made nothing that was made. In him was life, and the life was the light of men. And the light shineth in darkness, and the darkness did not comprehend it.'

What I know is that there remains after millennia, still essentially only two possible answers for the question of life on earth. Either life was created or life is an accidental molecular event. Up until the last 150 years or so, almost all of mankind accepted that life was created by God or gods or something supernatural. Now, more and more people, particularly well educated people, believe that life is a result of evolution.

Life is created or life is not created. Either Or. For one to be true the other must be false. Which is true?

One or the other. The stakes are very high. Winner takes all. There is no question more important. The answer determines our entire lives and the laws we live by because those who get to tell the creation story rule society. They get to define the nature of life and thus mandate how we live it.

If Darwin is right and life is an uncreated accident, then humans simply have to decide for themselves whether or how to survive their meaningless lives until their meaningless deaths. If however life is created it has a Creator.

The confidence with which almost everyone asserts Evolution as a fact is obvious and widespread. But that confidence disappears when asked about how it started. Evolution can only operate if something is already alive and able to mutate. Evolution is a mechanism, a process which causes living things to change. But something living has to kick it all off. But Science has not been able to get to that basic first starting block. Its absence is not simply a missing piece of information which it would be nice to have, a curiosity to satisfy and, a box to finally tick. The missing origin of life answer is a festering irritation because on it the entire edifice of third millennium life sciences, culture, philosophy and social policy has been constructed.

All of life on earth is believed to be explained by this one big idea, but it is an idea which is missing its crucial first premise. The problem for the big idea of Evolution is that the missing piece, the origin of life piece, is not just a missing link, it is the entire foundation on which evolution rests. On it everything else stands or falls. And yet it is not there.

I would have thought that as science advances in discovering the properties of living cells the problem of the origin of life would diminish. Instead exactly the opposite has happened. The problem in explaining the origin of life is that even the simplest living organism, a single cell, is unexpectedly and unbelievably more complex and sophisticated than ever anticipated. The magnitude of the problem just keeps getting bigger and the horizon in the quest to scientifically explain the origin of a single first cell just keeps moving further and further away.

Believed to be one of the best known origin of life researchers for the past 20 years, biochemist Klaus Dose states that, 'More than 30 years of experimentation on the origin of life in the fields of chemical and molecular evolution have led to a better perception of the immensity of the problem of the origin of life on Earth rather than to its solution. At present all discussions on principal theories and experiments in the field either end in stalemate or in a confession of ignorance.'[1] Even the most famous evolutionary biologist, Richard Dawkins, who wrote a bestseller called 'The God Delusion', publicly admits, 'Nobody knows how life began.' He must be oblivious to how illogical, scientifically speaking, his own words are when he says, 'We have *no evidence* about what the first step in making life was, but we do know the kind of step *it must have been*. It *must have been whatever it took* to get natural selection started. Before that first step, the sorts of improvement that only natural selection can achieve were impossible. And that means the key step was the rising, by some process as yet unknown, of a self-replicating entity.'[2]

From textbooks and prominent scientists alike, phrases like Dawkins' proliferate. I kept finding statements conspicuously vague. Not what I expected to find in science textbooks. Not what I understood as facts. Here are some: how genes *may have* originated... nucleic acids *may have existed... somehow,* by *processes yet unknown* genes evolved... metabolism *may have* been refined... the emergence of protobionts *may seem unlikely*... molecular biology today *may have* been preceded by... further changes *would have been* possible... self-replicating molecules and a metabolism-like source of the building blocks *must have* appeared together... Necessary conditions *may have been*...

It seems to be a matter of faith that abiogenesis did occur and that each scientist can, 'choose whatever method happens to suit him personally; the evidence for what did happen is not available.'[3] Trying to assemble a scenario in which life is an outcome of 'chemical evolution' and, what Paul Davies calls 'spontaneous complexification'[4] has so far been nothing more than many inadequate cases of wishful thinking.

I understand that there are countless unanswered questions about life. What is bewildering, therefore, is not that textbook science has no evidence of the chemical origin of life, but that it illogically insists, in spite of its acknowledged ignorance, that it *must have* been a mechanistic material process. Not only does it declare that it knows how it must have happened when it doesn't, it also fosters the illusion in the general public that it does. And it persists in this adamant obstinacy in the face of more than a century of failed scientific experimentation in creating *any type of life whatsoever, no matter how basic*. Evolution is a mighty ideological palace built on sand.

Evolution can never command complete confidence and intellectual conformity until the missing first piece is found. Science is betting on finding a piece that reinforces what it already largely believes. It wants an answer that it can swallow, one that confirms that life is uncreated, one that eliminates a creator God, one that doesn't undermine the entire scientific cultural atheistic paradigm that has grown out of Darwinian evolution.

(1) Dose, K., 'The Origin of Life: More Questions Than Answers' Interdisciplinary Science Reviews (13:4, 1988) 348

(2) Dawkins, R., The Greatest Show on Earth: The Evidence for Evolution (Free Press/Bantam, UK, 2009) 419

(3) Kerkut, G.A., *Implications of Evolution* (Pergamon Press, London, 1960) 150

(4) Davies, P., Lecture on the Origin of Life given on board the Beagle http://www.youtube.com/watch?v=jGald-E0BF4 viewed 12 April 2013

CHAPTER 23

THE MEGA-MILLION DOLLAR QUESTION

For nearly a century Darwin's theory provided the 'scientific' cover to banish a 'creator' in explaining the origin of life. Darwin thought a cell to be little more than a blob of jelly. But since the discovery of DNA and since significant technological advances in microscopy it has become less and less convincing. To support the theory that professes that something as mind blowingly intricate and sophisticated as a living cell could have come into existence by accident, resembles a facsimile of faith rather than science.

By 2005 the established scientific academy responded. A heavyweight stepped into the ring: Harvard University. It is the wealthiest, most powerful, most prestigious university in the most scientifically powerful nation in the world. Harvard knew that despite all the assurances given by biology textbooks the origin of life question was still the holy grail of all human questions. The name of Harvard would be, Harvard decided, forever acclaimed as settling the persistent nagging thorn in the side of Evolution. It would solve the question. And it would achieve its bold vision by giving over many millions of dollars to the greatest mystery of the scientific world; how did life begin?

Harvard established its very own 'Origin of Life Initiative' to search for evidence which would finally and incontestably settle the case. 'Our vision is to make Harvard the leader in origins of life research and education.'[1] By doing so Harvard openly acknowledged what I had suspected all along. 'The well-known conundrum of this emerging frontier is that we do not yet have a fundamental definition or understanding of life. Similarly we do not understand life's origins – how life emerges from chemistry.'[2] This amounts to probably the most authoritative confession the man on the street will ever get – that the textbook position of 'chemical evolution' is radically less conclusive than it pretends to be.

The Origin of Life in the Universe initiative held an open day for interested students to listen to keynote specialist scientists. A student posted her impressions of the day on the internet: 'It may be difficult to believe,' she wrote 'but there was a common theme to this seeming cacophony of scientific expertise and discovery. The theme was, "We just don't know." No one knows how life began - or even how to define 'life,' if you want to get all philosophical about it - but it's a question of such paramount interest and importance that key players in many avenues of scientific research are willing to devote their time and resources to answering it. Underneath it all, it was refreshing to hear a bunch of really smart folks say 'we don't know.' It was humbling and put things in a grandiose perspective. No one knows how we all got to be here, but the researchers in the Origins of Life initiative and beyond are trying to find out.'[3]

The Harvard strategy is to take a broad imaginative multi-pronged approach. Research into areas outside of the life sciences is encouraged.

One approach, astrobiology, involves the possibility of life coming from outer space. This theory of panspermia postulates that life, having originated beyond our planet, then 'seeded' itself between planets. 'After all,' says Harvard University, 'central to the question of the origin of life is the question – is there life elsewhere?'[4] Missions in space are planned to test whether cells embedded in rocks can survive an inter-planetary journey to earth. Scientists meanwhile continue to search for fossilized traces of life or actual cells in rocks which have entered the atmosphere from Mars.

Other lines of investigation being funded include the biochemistry of life as it is revealed in real cells on earth. For a few dollars you can buy all the ingredients to make a human being. In material terms humans are pretty cheap and basic: some water, a few lumps of coal, chalk, salt, a sprinkling of iron and trace elements, and that's about it. Around 99% of the human body is made up of oxygen, carbon, hydrogen, nitrogen, calcium, and phosphorus. What's missing of course is the recipe, the priceless information on how to put the ingredients together. Should anyone be able to assemble all the inert raw materials, what would be fabricated would be, scientists imagine, more than a corpse. It would be alive. Life, for Harvard scientists, is the result of finding the information regarding the arrangement of matter because as yet no scientist has ever known how to combine the ingredients and *make them live*. "My expectation is," said the team leader, Professor of Chemistry and Chemical Biology, David Liu, "that we will be able to reduce this to a very simple series of logical events that could have taken place with no divine intervention."[5]

Since the launch of the Harvard initiative no unicellular living organism has been engineered from inert raw materials. It's admittedly become ever more possible to cut and splice to make variant organisms and components of cells from living sources. It's technically possible to build Miller-Urey type experiments which mimic the early earth conditions far more accurately than the original experiment to see whether viable building blocks for proteins can be spontaneously assembled. But until now scientists have only ever been able to play with the parts. No-one has ever made, as in *created,* life. Life has only ever come from other life.

This law, that *all life* only comes from other existing life, is known as biogenesis. It was originally postulated by Louis Pasteur and continues to be unequivocally confirmed as a basic foundation of all third millennium life sciences. A universally applicable scientific law by definition admits no exceptions. Biogenesis is one such law.

What biologists call abiogenesis, life from *non-living* matter, has never occurred. But it is dogmatically expected, without any evidence it must be said, to have occurred at least once. Indeed it is even claimed to be a scientific fact based on the illogical, patently unscientific assertion that because it *had* to have happened, it *must* have happened. Life arising from a chance arrangement of molecules is a desperately needed theory for those unwilling to consider life being created. It is a way to dismiss divine intervention in creation. Atheist scientists rely on finding an original instance of life from non-life, a single aberration from the norm. They need evidence which contradicts their own law of biogenesis. The glaring irony here is that the scientific search for the origin of life lies outside the laws of science itself.

Another irony is inescapable. It has to do the three hundred year old motto of Harvard itself, 'Veritas' (Truth). The Origin of Life Initiative makes its claim to be seeking Truth questionable because only certain types of questions are allowed to be asked. Only certain types of answers are acceptable. No evidence outside of scientific materialism is admissible. No evidence of design will be tolerated. No ideas will be funded which venture outside of the paradigm of life as a mystifying meaningless arrangement of chemicals. 'Veritas' for Harvard is categorically compromised by a self-mutilating self-imposed ideology which rejects from the outset, anything other than chemical evolution. For Harvard University, Truth has already been decided, at least in kind, before it has even been found.

Another privately funded search in 2012 was no different. In the absence of notable progress emerging from the Harvard initiative, Harry Lonsdale, a millionaire chemist, launched his own 'Origin of Life Challenge: How did Life begin.' Once again the question was asked. How did Life begin? And once again it was a conditional question with a constricted answer because the competition was open only to proposals which address the 'conditions, materials, and energy sources believed to have existed on the pre-biotic Earth…and how primitive life could have evolved to modern biological cells.'[6]

Lonsdale said 'My goal in supporting Origin of Life research is to help scientists solve one of the great remaining problems in biology. A solution will give every science teacher in the world, from high school to college, a fundamental understanding of how life probably began on the Earth. In time, the world will find that the laws of chemistry and physics, and the principle of evolution by natural selection, are sufficient to explain life's origin.'[7]

He made no bones about the fact that the competition winners would be those scientists focused on life emerging from an RNA world. A panel of international scientists scrutinised the submissions from over seventy entrants vying for the $50,000 prize money and access to millions of research dollars. There were no independent researchers among the winners, all of whom were part of established institutions. When questioned about the quality of the proposals Lonsdale expressed his disappointment. "I must say I was very much underwhelmed by the breadth of their proposals."[8] In an interview he was challenged for his pre-commitment to Darwinism. 'Life' for the purposes of the competition, 'is defined as a self-sustained chemical system capable of undergoing Darwinian evolution.'[9]

Private philanthropists were not alone in funding origin of life research. NASA, it seems, has decided that on behalf of American taxpayers it should buy the right to erect an American flag on the pinnacle and glory of the scientific revolution. In September 2012 NASA awarded grants totalling 40 million dollars to five institutional research teams to study the origin, evolution and future of life in the universe. One of the recipients, Professor Nigel Goldenfeld, Swanlund Professor of Physics and leader of the Biocomplexity Research Team at the University of Illinois Institute for Genomic Biology will be using his 8 million dollar grant to explore, "How did life evolve in the pre-Darwinian era? There had to be much more rapid evolution from nothing to something. In about a billion years (from about 4 billion to 3 billion years ago) life evolved from nothing to something - and that's pretty fast. We want to understand how that could have happened, how life would have evolved before Earth had modern cells."[10]

What relief there will be for Science when it finally triumphs over the uncertainty of the origin of life? Who will get to claim that honour and be remembered forever in the annals of world history? Will it be Harvard? Or NASA? Or someone else? How many scientific questions will finally be laid to rest? Imagine the global jubilation if life from non- life can be finally, irrefutably, observed and scientifically documented.

Celebrations are in order. At last. The search for science's holy grail of abiogenesis is over. Life from non-living material has occurred. It has been observed, recorded, photographed, filmed and independently analysed in reputable laboratories by scientists from four continents. The evidence is real. It exists. It is seen. It is reproducible. Life has come from non-life. Not once, not twice, not three times, but four times.

(1) http://origins.harvard.edu/ viewed 12 April 2013

(2). http://origins.harvard.edu/ viewed 12 April 2013

(3) Posted by Anna Kushnir, *The Origins of Life on Earth*, Really on Mar 9, 2009 http://blogs.nature.com/boston/2009/03/09/the-origins-of-life-on-earth-really viewed 12 April 2013

(4) http://origins.harvard.edu/ viewed 12 April 2013

(5) Boston Sunday Globe, 14 August, Gareth Cooke http://www.darwinthenandnow.com/2012/05/race-for-the-origin-of-life-theory/ viewed 12 April 2013

(6) Mazur, S. Interview, "A Challenge to find Life's Origins" Astrobiology Magazine http://www.astrobio.net/index.php?option=com_retrospection&id=4778&task=detail viewed 11 March 2013

(7) http://www.originlife.org viewed 11 March 2013

(8) Mazur, S. Interview, "A Challenge to find Life's Origins" Astrobiology Magazine http://www.astrobio.net/index.php?option=com_retrospection&id=4778&task=detail viewed 11 March 2013

(9). http://www.originlife.org viewed 11 March 2013

(10) Des Garennes, C. 'NASA grant to UI aims at General Principles of Life' 18 September 2012 http://www.news-gazette.com/news/education/2012-09-18/nasa-grant-ui-aims-general-principles-life html viewed 12 April 2013

CHAPTER 24

EX NIHILO

The scientific community thus far has no concrete cases of life coming from non-life. Darwin saw neither his warm little pond nor his ancestral cell. This is what I clearly saw:

1. In four instances, in four places, Buenos Aires, Sokolka, Tixtla, and Leginica, extraordinary material, physical transformations had occurred. They could be and still can be, seen, touched, photographed and scientifically examined. They all still exist today in the 21 century. To the Roman Catholic Church they may permissibly be called miracles, because the term resonates in its milieu. But in secular scientific terms they are not miracles. They exist. They are not theories. They are unexplained absolutely physical and entirely material phenomena, available to scrutiny.

2. In all four cases there were many witnesses. By the standards assessing witness credibility in any court of Western jurisprudence they were excellent, reliable, credible and corroborated. What they unanimously assert is that the initial substance was bread which, of its own accord, transformed into something which was not bread, and which resembled blood.

3. In all four cases the transformation occurred over a matter of days, if not less.

4. All four cases were scientifically analysed by different reputable independent scientists in different reputable independent laboratories.

5. In all four cases an inanimate vegetable substance had spontaneously transformed into an animal substance.

6. In all four cases bread, fine wheat flour and water, had transformed without assistance or interference, into human cells. They were not tampered with in any way, a fact which was attested to by witnesses and then verified by the actual histological forensic analysis. It confirmed that no human fraud, no human ingenuity, no human manipulation or technology could possibly have created the results they found.

7. In all four cases the cells were identified as heart muscle tissue and human blood. Where tests for blood type were done, blood type was ascertained to be AB, which is the same blood type identified in the historical case of Lanciano which analysed a transformation that had occurred in the eight *century*.

8. In all four cases ample, fresh quantities of DNA were present.

9. In three cases, where DNA testing was done on ample quantities of nucleic material, there were unprecedented results: the curious inability to extract a genetic code.

10. In all four cases the accounts of what happened, the subsequent scientific analysis, and the final reports are public and transparent.

11. All four cases had their origin in Roman Catholic Churches.

12. All four cases involved material components, baked wheat flour and water, used in an ancient liturgical rite known as the Holy Sacrifice of the Mass.

13. In all four cases the four Bishops of the four dioceses, the men authorised by ordination as the primary teachers of truth (magisterium) in their jurisdiction, agreed to having the transformed hosts analysed by specialist scientists.

14. In all four cases scientific reports of what was discovered were issued to the Bishops of the four dioceses.

15. In all four cases the Bishops have accepted the results, which means they have accepted the fact that human cells have spontaneously emerged from non-living matter.

16. In all four cases Bishops of the Roman Catholic Church, one of whom, now residing in Rome, is the most famous Bishop in the world, have accepted that the human cells which emerged from non-living matter had no ancestral heritage. That is they formed spontaneously and without any biological parentage.

17. In all four cases the Bishops confirm, by their acceptance of the scientific findings and testimony of witnesses, that something has come from nothing. The constituent chemical and physical properties, that is, the complete range of material ingredients of a human cell, are not present in bread and water.

For the first time science has testified to an act of creation. A new creation has come into existence from nothing: ex nihilo

CHAPTER 25

FROM THE MOTHER

Blood had come from painted plaster. And blood had come from bread. These were two completely different cases and both were unlike anything science had seriously explored. From the tests that we had done on the blood encrustations from the Cochabamba statue of Christ and on the Eucharistic host of Buenos Aires, we had noticed that there were remarkable similarities:

1. The nuclear DNA tests from both confirmed the presence of *human* DNA.

2. The pathology tests from both showed evidence of inflammation associated with recent *traumatic injury*.

3. The nuclear DNA tests from both confirmed that the samples provided ample *quantities* of material to be tested.

4. In the opinion of the scientists involved with both cases the *quality* of the sampled material was high and would furnish results easily. And yet, in spite of high quality, high quantity, high expectations, and in spite of repeated attempts.

5. The nuclear DNA tests from both were uniquely, mysteriously, *unable to provide a genetic profile*.

Why not?

Why had many thousands of forensic DNA tests been conducted in criminal cases, paternity cases, historical cases, and provided a genetic profile of the identity of the individual but in both our cases, both identified as human, there was a blank?

There had to be an explanation for this same result coming up again and again. Yes we know there's DNA but we can't read it. Why was there no genetic profile of the human being whose DNA was in the nucleus in the cells in the blood in the samples from the host and the statue?

Some very basic knowledge of human biology explains that in the nucleus of every cell is the DNA which is unique to an individual. DNA is a very long molecule made out of billions of chemical bonds. When these chemical combinations are identified they reveal a specific unique code for each species, and each individual. Every human being has a unique genetic profile.

When a human male body makes sperm, each sperm cell has exactly half of the code. When a human female body makes eggs, each egg has exactly half of the code. So when a human male sperm cell fertilises a female egg the two halves of the parental DNA molecules fuse to form a whole new DNA molecule in the nucleus. A full human genetic profile comes from the combination of the genetic material. Half comes from the father and half from the mother through sexual union. A human embryo with an entirely new DNA code comes into existence at the moment of conception. That new DNA becomes the unique blueprint for the developing child. It is replicated trillions of times over to become the same code within the nucleus of every other cell of that growing child, be it a heart muscle cell, a bone cell or a white blood cell.

It occurred to me that if we had obtained a full human genetic profile in these standard DNA tests which tested the DNA in the nucleus then the blood in these cases could not possibly be the blood of Jesus Christ. Jesus after all, it is dogmatically believed, *was not the product of a sexual union.* He had no biological father. Therefore any profiling of the DNA of Jesus would not have standard results. There would be no standard DNA with genetic material from a biological father.

Recently however scientists have discovered another source of DNA. It is also inside every cell, but it is not in the nucleus of every cell. It is found outside of the nucleus in the organelles called mitochondria. Mitochondria are like generators in the cell. They convert energy for use by the body.

Analysing mitochondrial DNA is a relatively new advance in forensic DNA identification. It has opened a whole new chapter in the forensics of human identification because of its differences from the DNA in the nucleus. Two of those differences are particularly noteworthy.

The first difference is structural: the nucleus is a bit like the yolk in an egg, soft and easily damaged. The mitochondria however are like little hard nuts, tough and protected on the outside. Because mitochondria themselves are structurally strong they protect the DNA within them. And because of this mitochondrial DNA is generally less degraded after many years than nuclear DNA. Mitochondrial DNA is therefore better suited to the analysis of *ancient* sources. Hair, bones and teeth can all produce a satisfactory profile.

Mitochondrial DNA analysis has successfully identified victims of mass natural disasters where the nuclear DNA in the cells was too badly damaged. The remains of soldiers from World War I who died on foreign battlefields have been identified and returned to their homeland. It has also been used in high-profile criminal cases such as the Boston Strangler, the Green River murders, and the Laci Peterson homicide. Mitochondrial DNA tests have been consulted in identifying the remains of the family of Tsar Nicholas II, the Italian poet Petrarch, King Richard III, St Luke, the writer of the gospel, and, in a twist of poetic justice, the father of modern science himself, Copernicus. In 2009 hairs found in a calendar owned by Copernicus were analysed and genetically matched with the mitochondrial DNA of the skeletal remains of a female relative buried in the Cathedral of Frombork in Poland.

The second difference concerns the origin of the DNA material in the nucleus versus the origin of the DNA material in the mitochondria. Nuclear DNA comes from the father and the mother. But mitochondrial DNA is passed on *only from the mother.* It is the mother's DNA profile. Grandmothers and grandchildren, mothers and children, all share identical mitochondrial DNA. Maternity and maternal lineage is identifiable. Paternity is not.

Jesus did not have a human father. But he did have a mother. And a mother gives to her child mitochondrial DNA which can be profiled.

And so it dawned on me that the logical next step in the process of examination of the statue and the Argentina host, was to test them for Mitochondrial DNA. Could we find the *mother's genetic code?* If there was a mitochondrial DNA match in these two diverse cases, one from a host and one from a statue, it would point to them being from the same maternal line. And if this was the blood of a human called Jesus Christ then the mitochondrial DNA would be from his human mother, the Virgin Mary. Jesus' maternal lineage would be recorded in a DNA profile of the mitochondrial DNA in our cases.

CHAPTER 26

CONTAMINATION IN CHANNEL 7

Mitochondrial DNA was an intriguing new avenue for testing. Mike Willesee and I planned to document our new mission to test for mitochondrial DNA, not only in the Cochabamba and Argentina cases, but also in some other recent cases of proclaimed Eucharistic miracles which had since arisen in different parts of the world. We thought that if we could establish a mitochondrial DNA profile connection in all or any of these cases it would place us in a position to compare our findings with future mitochondrial DNA testing on blood on the Shroud of Turin.

We tentatively titled our proposed documentary 'The Blood of Christ' anticipating global fascination in a story which scientifically probed the genetic make up of the most unusual, most influential person in the history of the world. It was hard to imagine a greater subject for a documentary.

We had already commenced work on this new venture when in May 2015 Channel 7, an Australian Television Network became aware of what we were up to. At that time Mike was working as one of their presenters on its weekly prime time television program, Sunday Night. Seven had meetings with Mike and me about cooperating on a television special which would be targeted for local and international distribution. The proposal was that 7 would revisit the work we had already done and then continue the story by taking on the new mitochondrial DNA testing. Seven saw great potential in the story and decided to put a lot of money into the venture. It was touted as the network's largest budgeted program of the year. In fact it would be its 'program of the year'.

One of Seven's senior producers, Alex Garipoli was assigned to produce the program. From our very first meeting Alex was full of enthusiasm. He wrote about it to me: "It was very good to spend time today getting my head around this fantastic story. My understanding is certainly heading in the right direction and I am really looking forward to commencing this journey!…"

In the next few months Alex viewed and downloaded many, many hours of my original video tapes; twenty years of my journeys to unusual destinations to record interviews and witnesses and scientific reports from various cases. What I had already recorded in documentaries and in my books *Reason to Believe*, and particularly, *Unseen*, became key reference material for Alex to understand the story, the key players and where our scientific research was heading. In August 2015 the Seven Network launched into a one month long overseas filming trip.

They wanted great visuals that would attract a broad secular audience. Dramatic intrigue, they surmised, would arise by anticipating DNA test results from laboratories. A leading Los Angeles cinematographer was enlisted along with truckloads of the latest film equipment including cameras mounted on drones.

We began in Buenos Aires with spectacular aerial shots over the city of former Archbishop Jorge Bergoglio, the man who had commissioned my very first work on the Buenos Aires Eucharistic miracle case. With contemporary filming technologies they sought to re-enact and bring to life what happened in that parish church in 1996 when the communion host had transformed. Seven was not able to obtain any new samples of the transformed host for DNA testing, and was relying on me providing a sample that I had obtained back in 2005, which could be the subject of testing for the program.

Padre Alejandro, who was central to the story, had been moved to Salta, a poor remote village in the mountains of Northern Argentina. There the camera crew commandeered a local restaurant overlooking the town square for the interview. Mike did one of his classic interviews. He elicited from the humble softly spoken priest not only the relevant facts, but the emotion that was involved in the remarkable events that had taken place in his church in 1996.

From Argentina we went to Bolivia to revisit the statue story. A new cat scan was filmed at the same laboratory in Cochabamba. The doctors there remarked, "We x-rayed this statue 20 years ago and found nothing. We x-rayed it again today, and nothing has changed." There was still no sign of human manipulation and fraud.

The statue's owners also agreed to Channel 7 taking new 'blood' samples from the statue so that they could be subjected to DNA analysis. A local pathologist was engaged to take the new samples. He delicately collected several pieces of dried blood from the face and forehead regions of the statue and inserted them into a glass tubes which were scrupulously labelled and sealed. The whole process was scrutinised on film by Seven and by me.

We also filmed in a remote village in Colombia and then went to Tixtla in Mexico to film the places and the events surrounding two other recent cases of believed Eucharistic Miracles.

Upon returning to Australia Alex approached the Victorian Institute of Forensic Medicine in Melbourne to do DNA tests on the samples we provided. I was hesitant because I was really concerned about getting a contaminated result. I had been researching if what I was hoping for was even possible. Was there laboratory equipment so sensitive that it could select and then isolate a single cell? Could that *single* cell then be tested for the DNA in it, and in it alone? My thinking went like this: Contamination of the samples was very possible because the statue had been touched and handled by so many people in the last twenty years. DNA from some unknown person's sweat or tears or skin could be on the bloody crusts we were about to test. If only we could make sure it was a white blood cell we were testing and only a white *blood* cell. Or only a heart muscle cell? I said to Mike and Alex that I thought the only way we were going to be able to know for certain whose blood it is in these cases is to find a lab that uses technology that can isolate the white blood cells in the sample, extract them and then profile the DNA from those cells, and *only* those cells. This way you eliminate any DNA that might be coming from other types of cells that can end up in the sample.

The Victorian Institute didn't do single cell analysis. They analysed samples comprising many thousands of cells which were entirely destroyed in the process. I had found another laboratory which did. Using Laser Capture Micro-dissection equipment the University of Melbourne could extract a single cell from an histology slide and transport it to a sterilized environment for DNA testing. But Channel Seven was adamant. My urging to go this route fell on deaf ears.

It felt to me that I was being sidelined from the work I had been doing for twenty five years, work on DNA and testing for it which had made me far more experienced than a television production team. I knew how easy it was to contaminate samples. The only way, I argued, to get real results is to make absolutely sure that the lab is testing for the DNA of a blood cell only, and not inadvertently testing for the DNA in saliva (from kissing the statue), or from a flake of skin (from handling and touching the statue.) But to them I was just the small guy who should be quiet and leave it to the professionals. They had a deadline. Following my suggestion to find another laboratory would mean more precious time lost. So Alex and Mike decided to go with the Victorian Institute of Forensic Medicine. It would be a decision that Mike would sorely regret as he revealed when I later interviewed him just before he died.

I had done what I could do. Precautions had been taken to try and avoid any DNA contamination. All of us had saliva swabs taken so that our DNA information could be recorded just in case one of us had inadvertently contaminated the sample and skewed the results.

The Statue sample yielded a full nuclear and mitochondrial human DNA profile. *It was from a woman.*

We were all deflated. The result made the bleeding statue look like we had been duped. The whole thing was a big fraud.

But since this was exactly what I had feared, I knew what to do next. Eliminate the prime suspect, the owner of the statue, Silvia Arevola. I had details of her nuclear DNA genetic code from testing that had been undertaken previously. I provided that information to the Victorian Institute of Forensic Medicine. Sure enough, it was a match. They had picked up DNA from Silvia's cells on the statue.

They hadn't managed to isolate the blood cells.

The sample from Buenos Aires bleeding host was only tested for mitochondrial DNA. The entire sample was consumed in the analysis. The material analysed yielded a mitochondrial DNA profile. However, this profile consisted of a mixture of at least two individuals and, the report said it "cannot be interpreted any further." One of the individuals whose mitochondrial DNA profile ended up in the results was a member of the film crew. This showed how easy it was for a sample to become contaminated. Simply being within breathing distance of the sample in a sterile environment was enough to contaminate the result. Once again the failure to isolate and test individual cells had been the undoing of the result.

My long-standing anxiety concerning contamination had played out. I had feared that external contamination could be an issue and it was.

With the results we now had, we were at another dead end. But if there was one thing I had learned in doing this work it was that whenever it seemed the road has come to end a new and completely unexpected one emerged. This time was no different. In very low spirits I wrote to the University of Melbourne asking if they could isolate and profile the mitochondrial DNA of a single cell. And once again an answer arrived by coincidence.

'There is some movement in this area of late" wrote Dr Dadna Hartmann from the Victorian Institute of Forensic Medicine. She had, serendipitously, just returned the week before from an international conference where a *whole new technology* was presented. They showed, wrote Dr Hartmann, "separation of different cells from crime scene samples, as well as *single cells isolated* from a mixed sample… with the single cells yielding a DNA profile. This work was done by a team in Italy…"

This time I did manage to convince the Channel Seven producers, and within weeks the whole team was on its way to Bologna's, Menarini Silicon Biosystems, a biotech company which advertised itself with a catchy slogan: "Every cell has a story to tell."

If, I thought, they only knew how true that was.

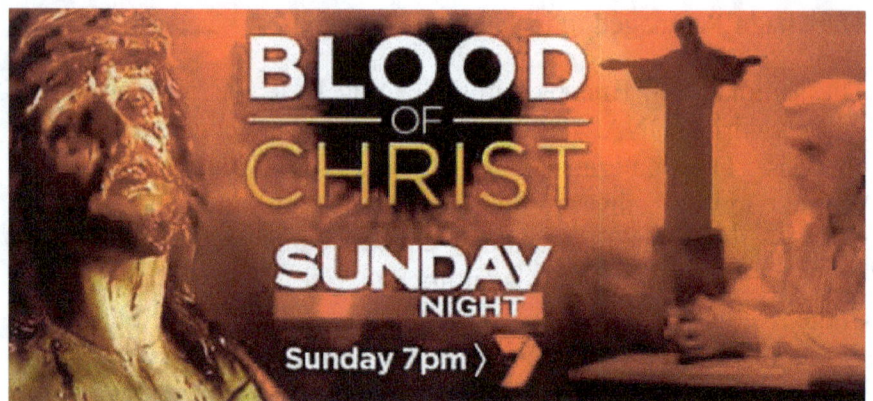

On Sunday 9 April 2017 Blood of Christ was broadcast throughout Australia.

On 27 August 2015, filming of the Buenos Aires segment began.

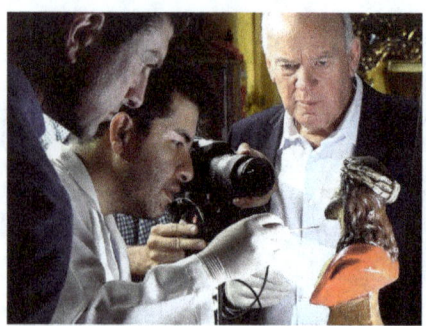 

On 4 September 2015 in Cochabamba Bolivia, a sample was taken from the forehead of the statue.

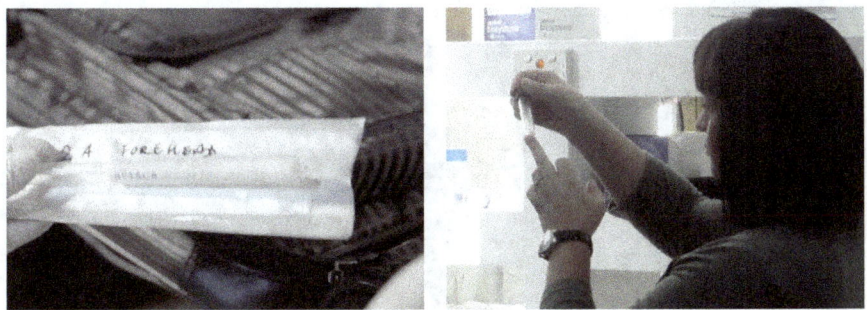

On 9 October 2015, one sample was tested by the Victorian Institute of Forensic Medicine. A genetic profile was obtained but was proven to be a result of contamination.

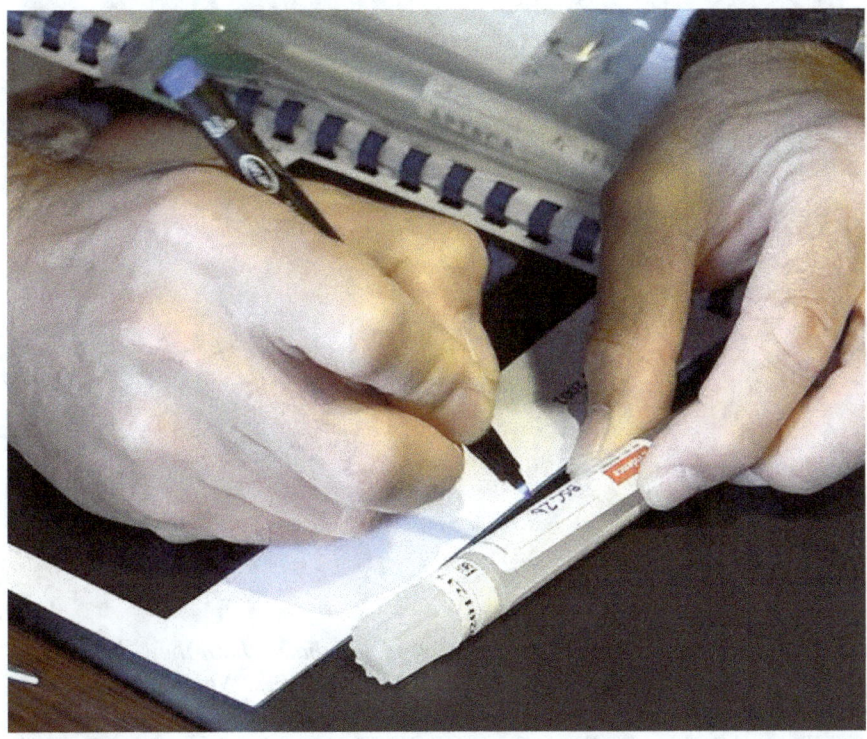

Another sample taken from the forehead of the statue was submitted for DNA testing with Silicon Biosystems in Bologna Italy, along with a sample of the 1996 Buenos Aires Communion host . This Buenos Aires sample was taken and held by me on 9 December 2005 and released to Seven for the Silicon Biosystems tests.

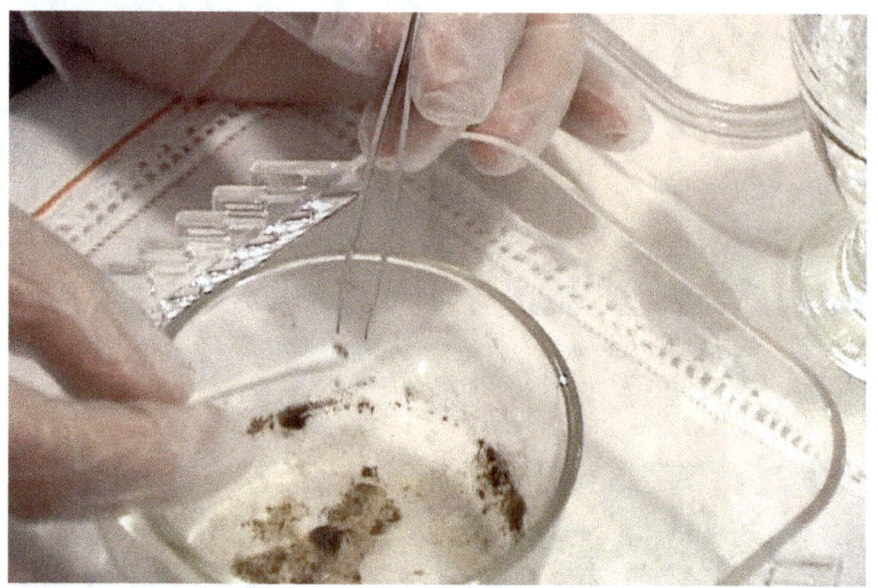

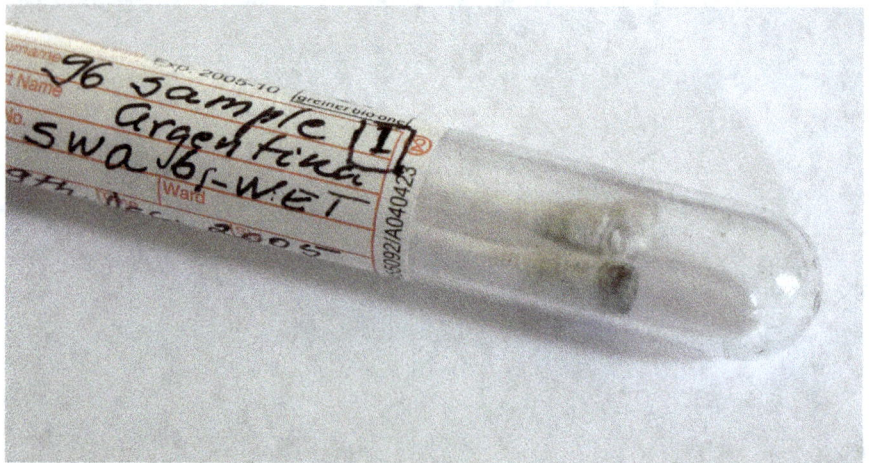

On 16 March 2016, Silicon Biosystems reported that in both the Cochabamba case and the Buenos Aires case there were similarities. No DNA profile was obtained from the nuclei. Mitochondrial DNA testing for maternal inheritance on single white blood cells however revealed **"a surprising amount of genetic data...We got unprecedented results in the history of Forensic DNA testing... which showed that "the cases were somehow related".**

CHAPTER 27

THE ITALIAN JOB

The stories that Menarini Silicon Biosystems are used to telling come typically from crime scenes. Cell samples are collected which, more often than not, are mixed, meaning for example, blood cells mixed with skin cells mixed with sperm cells. The difficulty in the past was determining who contributed what cells. So the Italian scientists at Menarini created a breakthrough technology called the Deparray system which could first identify, and then separate the cells from each other, and then genetically analyse each one of them separately. Each of these processes is an astounding achievement. But to do all three, "with 100 percent accuracy", is a monumental step forward for DNA analysis.

So far in my 20 years of trying to find out for certain whose blood was on the Cochabamba statue or whose tissue cells were in the Buenos Aires Eucharistic host I have had to rely on technology that was not capable of distinguishing whether any DNA was coming from blood cells or from contamination through someone else's skin or saliva cells. This brand spanking new technology had come just at the right time. I couldn't but see it as yet another remarkable solution to a seemingly insurmountable obstacle which lay in the path of this work of my lifetime. I couldn't see it any other way but as 'God sent.'

In late February 2016 I travelled to Italy with the Channel Seven producer and, in my hand luggage, one of the precious samples from Buenos Aires that was collected on 9th December 2005. I had not let it out of my possession since the time of collection and yet here I was, in my hotel room in Bologna, filming the sample being formally transferred into the custody of the producer. He alone would be walking into the Menarini Laboratory to request the Deparray analysis. I was to remain in my hotel room, entirely anonymous.

I had by then learned a hard lesson which I had to explain to Channel 7. That although science enjoys an authority which rests on objective facts and the scientific method, scientists themselves are not themselves always pure seekers after objective truth. Some scientists are not content to simply follow the evidence wherever it may lead. Some scientists don't like, or don't want to find, results which contradict their own ideology. Some scientists don't want to jeopardise their sources of funding or their jobs. So in order to try and keep ideology out of the picture I had to remain out of the picture. For the focus to remain strictly and entirely on pure scientific testing I had to be completely anonymous. It was too easy to do a Google search and find me associated with more than twenty years of research into strange events occurring in Catholic churches. The mere word 'catholic' was enough to set them against me.

I explained, by way of example, the encounter I had in Germany, with an esteemed scientist, Dr Susanne Hummel. She had been recommended to me as an expert in DNA research and the author of a definitive textbook on DNA typing by Dr Tom Loy from Queensland. Dr Loy suggested she might be able to extract and profile the DNA in the sample.

Just before the producer left the hotel to go to the lab I filmed the samples that were about to be tested and also my formal transfer of custody of the sample from the Buenos Aires case that I had in my possession. The samples from the statue were new, taken by the Channel 7 team on our recent trip filming in Cochabamba.

The plan was that the Italian lab would first find and separate white blood cells and epithelial cells, and then use some of each type of cells for standard nuclear DNA testing. Nuclear DNA testing routinely yields the genetic code of a person, a code which is contributed half from the father and half from the mother.

From the Buenos Aires sample 42 cells were recovered and used for nuclear DNA testing. From the Cochabamba sample they recovered and tested 18 cells.

Menarini Silicon Biosystems reported back that they didn't understand why, that it was 'not possible to understand why,' but they were *unable to generate a human genetic profile* from any of those cells.

Their confidence seemed shaken because they had assured Channel Seven that the kit they were using was Power Plex FUSION 6 recognized to be the most complete one, with its 27 loci. This kit had been tested by themselves, they wrote, and by the carabinieri (the Italian police) with single cell input, and it performed excellently.

For Alex and the Seven network this was a huge disappointment, a big nothing which they couldn't use. But for me, it was pure validation and highly, highly significant. This was the first time that the critical white blood cells themselves, and white blood cells alone, were ever isolated and tested. The no-result was a major result. It meant that the person whose blood it was had no discernible human genetic code. It confirmed my prior nuclear DNA testing in different laboratories, at a time when the samples were fresher and pathology analysis showed that the cells were in good shape.

Menarini Silicon Biosystems then went on to apply Mitochondrial DNA analysis to our samples in an attempt to obtain the maternal ancestral code.

The technology used to identify the types of cells was similar to facial recognition technology. A computer program holds information about the standard shape and morphology of the relevant cell types and then scans the sample in a sophisticated recognition program to make its identification.

Menarini Silicon Biosystems reported back on how well their Deprray system performed retrieving and identifying the cells. However they drew our attention to something odd about the appearance of the white blood cells and wrote that,

"the morphology (the size and shape) and the staining of white blood cells *did not look like what we typically see...* The dimensions were similar to regular white blood cells and the morphology close to, but not quite identical".

They nevertheless proceeded testing the cells that were selected with Next Generation Sequencing (NGS), the latest, most advanced technology that can sequence the entire human genome. "For sequencing, we used a NGS approach: we used the Precision-ID mtDNA Control Region panel kit (Thermofisher) for library preparation followed by IonTorrent PGM sequencing."

Because we were seeking genetic information on the matriarchal lineage only, the non-coding region of the human genome was selected for testing. Waiting for the results was tense for everyone because the program was already being assembled and the editors and producers were awaiting the final piece of the story. The deadline for broadcast was fast approaching.

We were hopeful and optimistic that the Italian lab would find that the Mitochondrial DNA tests results were the same for both the Cochabamba statue case and the Buenos Aires host case. If the results were the same we could infer that we had found the genetic markers for the mother of Christ. The question that had been put to Menarini Silicon Biosystems was " is there a DNA match between the two cases being considered.?"

Then their report came in.

'We were able to identify, and to isolate, a group of pure cells, such as, epithelial cells, and, white blood cells, out of each sample. And for each group of cells, we were able to get good mitochondrial data. We were surprised at how much genetic data we obtained.' They went on to say, *"We got unprecedented results in the history of Forensic DNA testing'* and that 'the samples are somehow kind of *related*'. The mitochondrial data obtained from the cells (both white blood cells and epithelial cells) in both the Cochabamba case and the Buenos Aires cases were in *some ways similar, but had differences which they could not interpret*. The differences, it was explained related to the fact that there was evidence of *heteroplasmy*. Heteroplasmy is defined as a condition where more than one Mitochondrial DNA sequence is found in one individual. The "heteroplasmy" factor had complicated the situation and it precluded Silicon Biosystems from saying there was an exact match. And so in the end the best they were able to say was "The samples are *somehow* related."

Had I been present when Channel 7 filmed the Menarini scientist delivering the results, instead of carefully hiding my association to the case by remaining in the hotel room, I would have immediately asked for further explanation. What do you mean by 'somehow related'? What exactly was 'unprecedented in the history of forensic DNA testing'? But no-one asked. And no-one was willing to brave the elite world of scientific jargon and assumption to seek a full understanding and interpretation of what was ostensibly pages of oblique code and numbers.

To properly interpret the heteroplasmy and how it might help or hinder a comparison of our two cases required specialist *expertise in interpretation* which neither Silicon Biosystems or the Victorian Institute of Forensic Medicine (VIFM) had. We were in conversation essentially with scientific technicians, not analysts .VIFM suggested that we approach Professor Walter Parson at the Innsbruck Medical University in Austria. He was a leading expert in Mitochondrial DNA and heteroplasmy. He was our man.

But Seven were not prepared to go further with the investigation. I asked them to approach Professor Parsons to have the results properly interpreted. I had thought that two very significant comments made by Silicon Biosystems, *'We got unprecedented results in the history of Forensic DNA testing'* and that 'the samples are somehow kind of *related*' would have been worthy matters to pursue in any genuine attempt to reveal the truth of this story. But Seven did not quite grasp the magnitude of the data at their fingertips. They were not interested.

They had, in their view, come to the end of the road with the ultimate in high tech testing and there was no big headline ending to the story. The failure of the Menarini Silicon Biosystems to be able to give a clear interpretation of the results took most of the gloss off the story and the program. They wanted to finish it and be done with an escalating budget by slashing the running time in half and terminating Mike's contract. They took it upon themselves to concoct a closing statement and shove the whole thing into the too hard basket.

Mike and I were given the opportunity to see their final edit of the program a few weeks before it was to be broadcast.

I was shocked by the way my film material had been selectively edited to present a diametrically opposite point of view from the actual truth. Simple facts and scientific information were manipulated to lead the viewer to conclude that both the Buenos Aires case and the Cochabamba case were suspicious, if not outright fraudulent. For instance, the DNA on the statue was proclaimed to be from a woman. This contradicted the truth that after the Italian lab did single cell analysis no human genetic profile was detected. *There was no profile at all, let alone that of a woman.* Seven studiously avoided the many genetic tests I had done over many years that said exactly the same thing, that there was this very strange thing going on, consilience, in that there was *no genetic code in the DNA*.

One sure way of confirming that the blood from the statue was not that of it's owner, Silvia Arevola, was to submit a sample of Silvia's blood to Menarini Silican Biosystems and for them to do precisely the same mitochondrial DNA testing as was done on the statue blood to see if there was a match. I had implored Seven to do this. They refused. So I did.

I arranged for Silvia's blood to be tested for mitochondrial DNA by the Australian Genome Research Authority using the same testing procedures and protocols used by the Italian lab when they tested blood from the statue submitted by Channel Seven. The results were that the mitochondrial DNA information that came from Silvia's blood did not match the mitochondrial DNA obtained from the blood on the statue. There was no match. It meant for certain, absolutely and categorically, that the blood on the statue was not that of Silvia.

For Seven to complete the program it needed access to much of what I had filmed over twenty years: events, scientific examinations, photos, interviews, and so on. I had provided this, and my time and effort, free of charge. However I made it a condition that my material not be used in a derogatory way. It had to be given respect and be presented fairly.

What Mike and I saw in the pre-broadcast screening started out well. The first three quarters were good and reflected the case histories well and Mike Willesee at his best. Both of us knew that this would be the last television work he would ever be involved in. His cancer was too advanced by then. This would be his finale; the final act in a long career as one of Australia's top journalists. The last part of the program however was unfair, erroneous and deceitful. It was enough to kill the truth.

Mike conceded that, "In the last few minutes it all goes wrong."

His voiceover states that we found the DNA of a woman in a sample from a Eucharistic host. But this was flat wrong. The sample had come from the statue. And later found to be contaminated by Silvia. He says to the camera, "So it's not the result of a divine event. It even suggests there may have been a well-orchestrated fraud."

Mike by then was much weakened in mind and body by the advancing cancer and he had lost all the star power he once enjoyed. He no longer had any bargaining power or influence in the direction of the final cut. He couldn't muster a defence for the truth.

I could and I did. With only days to go before the public broadcast I wrote a letter, as a lawyer, to Seven setting out my objections and threatening to take injunction proceedings to prevent the program from being broadcast if the program was not properly corrected.

I was extremely disappointed because I had put so much time and effort into not only the making of the program, but into years of research and my own documentaries that preceded it. The reputation of all that work was on the line too. I had an enormous burden on my shoulders. Either I said no, and the program never went to air, or I said yes to a half-hearted postscript tagged on feebly at the end. Everybody knows that once false, incomplete or unsubstantiated information is promulgated it's almost impossible to correct with truth, because first impressions are so powerful.

My instincts were strongly to refuse. But Mike, my old friend and associate in twenty years of investigations, was rapidly approaching his end, undergoing chemo and immunotherapy, and it weighed on me that I would be the instrument to obliterate his very last piece of work. Either way I was unhappy. Seven was under no circumstances prepared to re-edit the whole thing. All they were prepared to concede was a feeble post script at the end of the program. I relented and gave in.

This is what the rushed token statement at the end said:

"And a special thanks to Ron Tesoriero, author and lawyer, for access to his astounding footage used in the story. And there's a further update in this on-going investigation. Tests have confirmed the genetic material we obtained from the Statue of Cochabamba contained DNA that matched that of the statue's owner, Sylvia Arebelo. Scientists concluded it's because she regularly handles the statue. But, in the Buenos Aires case, numerous tests have failed to identify whose blood or heart tissue was found on the communion host."

The show was broadcast to great fanfare but predictably fizzled out to the ignominy of being forgotten. There was no big reveal at the end, only confusion and deflation. I was at a very low point. But even then, after so many years of pushing through and over obstacles, I clung to my conviction throughout the many years that I had devoted to this work: that help, Divine help, would always be forthcoming. Eventually.

CHAPTER 28

ANSWERS IN ARAMAIC

The written word for 'heart' in Aramaic is achieved by combining two ideographs, two picture symbols. The first is a shepherd's staff, symbolizing 'benign authority', the very same symbol used in the word for 'God.' The second is a floorplan of a tent, symbolizing 'being inside'. 'Heart' is 'the authority inside'. Unsurprisingly there is no Aramaic word for 'conscience'.

If the Buenos Aires and Cochabamba cases were genuine repositories of the blood of Christ, as Mike and I had believed, then there should have been a maternal DNA match. Jesus had a mother. There had to be a shared maternal genetic identity. But there wasn't. What went wrong? Why was there no match? Were the Menarini Lab results correct?

Deflated after the collaboration with Channel 7 and not knowing how to move forward I was tempted to just give up. It was all too hard. In my prayers I asked "Where do I go from here, Lord?" I remembered the early days when I first embarked on the study of the Buenos Aires host case in 1999. Katya Rivas had given me a message from Jesus: "Ron, Do not be concerned if you come across contradictions or setbacks if I am going to be guiding you and your interests are mine."

In that instant I made up my mind. I decided to fly to Bolivia and ask Katya Rivas if she would intercede with Jesus on my behalf. I wanted to know what to make of the Menarini lab results. Mike agreed to accompany me. It was a hasty decision, irrational perhaps, and profligate to some who would be puzzled why I would spend thousands of dollars to travel to the other side of the planet and return after two days. But for Mike and me there was a lot at stake and we knew from past experience that when an important matter cropped up in the work we were doing, Jesus, through the intercession of Katya, did sometimes give directions which set us on the right path. We needed help in the situation we were faced with now.

So within days I was sitting at Katya's dining room table with her faithful spiritual advisor, Father Renzo, a nun who was interpreting, and her daughter, Tatiana. Mike was resting in an adjoining room. I played her the Channel 7 documentary on my laptop. I explained what had happened with the mtDNA testing in Italy, about how detrimental the DNA results from Bologna were for the Eucharistic host case and for the Cochabamba statue case.

I asked, "Can you ask Jesus, what went wrong, and why did we not get a perfect DNA match?"

Katya paused and in a calm and thoughtful manner replied, "Jesus said the white blood cells did not have their usual shape. The reason is that white blood cells change their shape when the body undergoes trauma and injury. Those white blood cells are showing what I suffered in My Passion."

I hurriedly scribbled down in my note book what she had just told me.

Me: 'Is he telling you any more?"

She did not respond. I prompted her again.

Katya: "Jesus is not telling me anything else."

Me: "I don't understand how what he said has to do with the mitochondrial DNA tests results."

Katya: "I don't know. But he is not saying anymore to me."

I prompted her a few more times in desperation. Eventually she replied.

Katya: "Jesus has given you the answer and you must work with that."

Later in the evening at 9.47 pm, Katya picked up her cell phone and noticed that some text had appeared on the screen. The first line was the date and time, '17 de Marzo de 2017 21.47' and was followed by text in an unusual script. I thought at first it was Arabic. Unbidden I immediately pressed the record button on my camera. Tatiana translated as her mother spoke.

Katya: This is a message from Jesus. It is in Aramaic.

Me: How do you know it is in Aramaic and that it is from Jesus?

Katya: "This has happened to me before."

Me: "Can you read Aramaic?"

Katya: "No."

Me: "How do you know what it says?"

Katya: "Jesus tells me what it means."

At 21.49pm, two minutes after the Aramaic text was sent, more text appeared immediately below the Aramaic text and comprised nine lines of Spanish text in two paragraphs. Katya held up the screen of her phone to show me so that I could see and film the Spanish text that had just shown up.

Katya: "Jesus said to me that it is a citation from the Gospel of Luke."

Tatiana translated the Spanish message:

"And Joseph went up to Galilee from the city of Nazareth to Judea to the city of David called Bethlehem because it was the house where David and his family lived. There were some shepherds in the same region lodging there and keeping nightly watch over their flock."

Katya offered no explanation about what the citation meant or why the message came to her when it did. All she was aware of was that it came from that section of Luke's Gospel which addresses the circumstances just prior to the birth of Jesus when Joseph and Mary went to Bethlehem to comply with a decree issued by the Roman authorities that all people complete a census and enroll in their own town of origin.

The fact that the message arrived while I was present and that I saw it arrive prompted Katya to say that Jesus must have meant it for me for some reason. I could not make any sense of it nor how it might have anything to do with my questions about the mitochondrial DNA test results.

I then asked her about something which I had recorded nearly ten years ago when we were discussing the scientific research we had done on the blood. This is what was recorded at that time, 7 September 2009:

Katya: " Dile que no van a enconrar un código genético porque ante de ser hombre, ósea hombre encarnado, Él es Creador. *(Tell them, meaning Mike and I, they're not going to find a genetic code because before being a man, that is man incarnated, he was the creator.)*"

Father Esau Garcia translated Katya's message from Jesus and proffered an explanation.

Father Garcia: "Jesus is saying that he is the creator of humanity, he does not have any human genetic code because he is God. This is the Gospel of John. When Jesus is talking to the Jews and they say, you are not 50 years old and yet you claim to have seen Abraham. Jesus replied, 'Prior to Abraham I existed; I am'."

Now, a decade later, I asked her to elaborate.

Me: "Katya, when Jesus said to you that he did not have a human genetic code I had assumed that he was speaking only of a nuclear DNA code and I assumed that was because he did not have a biological father. But that's not what he actually said. He said we would not find a genetic code because he is the creator. It did not occur to me that he might also have been addressing that there is no mitochondrial code. But reflecting back now, I think I must have misinterpreted what he said. I think Jesus must have been telling you that he has no human genetic code at all. No nuclear DNA profile and no mitochondrial DNA profile. What do you think?"

Katya: "All I can remember is that Jesus told me that he has no human genetic code."

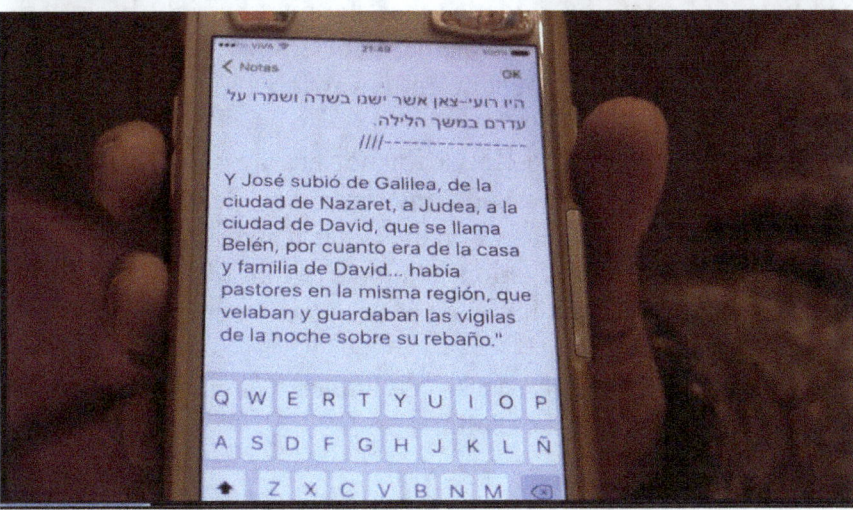

I asked a question about the Mitochondrial DNA results and his maternal ancestry. A message from Jesus, first in Aramaic and then in Spanish , appeared on Katya's phone pointing me to the answer. (17 March 2017 Cochabamba Bolivia)

I don't know if anyone could imagine how I felt when I heard that statement. I was privileged to be hearing something that has not been known ever before, not ever in the history of mankind. That Jesus had no genetic code. This was a profound completely new revelation about Jesus and one that could be potentially verified through further scientific testing.

Ever since cutting edge genetic technology has entered the field of determining ancestry there has been a fevered interest in the DNA of Christ. Projects involve the bones of crucifixion victims found in Jewish burial tombs, analysis of relics of John the Baptist, and the many stories of attempts to obtain the DNA of Christ from the Shroud of Turin. Investigators, like Mike and me, are seeking to use the newest scientific tools available to find answers.

It has been universally assumed that Jesus did have a human genetic code and that given time it would be discovered. No one before Katya has ever to my knowledge, suggested that Jesus never had a human genetic code at all. It was a monumental statement, one that could entirely discredit her, and her nearly twenty years of receiving messages from God as a secretary of Jesus. But I already had potential scientific evidence that she may be right because in the nuclear DNA testing of the samples from Buenos Aires and Cochabamba, no human genetic code was found.

The reason Jesus gave to Katya for not having a genetic code is that he is the Creator, he is God and that he existed before He was incarnated, before he became man. What Katya said about Jesus existing before his incarnation is supported in biblical texts. The bible says nothing about him having no genetic code.

The prior existence of Jesus, as Father Garcia explained, is mentioned in the Gospel of John,

57 Then the Jews said to Him, "You are not yet fifty years old, and have You seen Abraham?"

58 Jesus said to them, "Most assuredly, I say to you, before Abraham was, I AM."

59 Then they took up stones to throw at Him; but Jesus hid Himself and went out of the temple, going through the midst of them, and so passed by.

That Jesus is our Creator, is also addressed in the Gospel of John

"In the beginning was the Word, and the Word was with God, and the Word was God. He was in the beginning with God. All things were made through him, and without him was not any thing made that was made" (John 1:1–3)

If through Jesus all things were made then clearly he was the creator of man, and of the DNA which allows us to exist and reproduce.

Oddly, in seeking support for the notion that Jesus had no human genetic code, I had no help from the Bible. Instead, I found support in the scientific and political realm at the highest level. At the White House on June 26, 2000, President Clinton came to the podium and addressing the world media, congratulated those who had completed the momentous scientific milestone of sequencing the entire human genome for the first time.

President Clinton said,

"Today, the world is joining us here in the East Room…We are here to celebrate the completion of the first survey of the entire human genome. Without a doubt, this is the most important, most wondrous map ever produced by humankind. The moment we are here to witness was brought about through brilliant and painstaking work of scientists all over the world, including many men and women here today… we have pooled the combined wisdom of biology, chemistry, physics, engineering, mathematics and computer science; tapped the great strengths and insights of the public and private sectors. More than a thousand researchers across six nations have revealed nearly all three billion letters of our miraculous genetic code. I congratulate all of you on this stunning and humbling achievement.

Today's announcement represents more than just an epic-making triumph of science and reason. After all, when Galileo discovered he could use the tools of mathematics and mechanics to understand the motion of celestial bodies, he felt, in the words of one eminent researcher, "that he had learned the language in which God created the universe."

Today, we are learning the language in which God created life. We are gaining ever more awe for the complexity, the beauty, the wonder of God's most divine and sacred gift."[1]

Dr. Francis Collins, who directed the Human Genome Project, and is currently the Director of the National Institutes of Health in the USA, followed Clinton to the podium and said, "It's a happy day for the world. It is humbling for me and awe inspiring to realize that we have caught the first glimpse of our own instruction book, previously known only to God. That instruction book," Collins wrote in The Language of God, "was written in the DNA language by which God spoke life into being."

The statements made by both President Clinton and Dr Francis Collins are made from a perspective of seeing God standing outside of his creation but putting into his creation the blueprint for human life, the human genetic code.

There remains the question, that if Jesus stood outside of creation and did not have or need a human genetic code , how does one explain the mitochondrial DNA information that showed up in the testing of our samples.?

Seeking the answer to that question lead me to more surprises than I could ever have expected.

---

(1) http://www.presidency.ucsb.edu/ws/?pid=58701

CHAPTER 29

LEARNING NEW THINGS

I went home deflated because I wasn't consoled with an unambiguous answer to my questions. I still couldn't understand the dead end I had arrived at in the Menarini Lab results. And Katya's messages were opaque. There were essentially three messages to contend with.

There was a message about white blood cells changing their shape.

Then there was a totally unrelated Lucan text in Aramaic describing events leading to the birth of Christ. A message to me in response to scientific results was a bible quotation? In Aramaic? From St Luke who didn't even write in Aramaic but in Greek?

And then there was the statement, from Jesus, from God himself, that in spite of being a man, he had no genetic code.

I didn't know what to make of any of it except for one thing: That Jesus was, by referring to the cells in the samples from the host and the statue, confirming that they were indeed his. These were cells from the Jesus who was talking to me through Katya. These were cells from the same Jesus who died two thousand years ago. In spite of not being able to understand I was confident that I was on to something.

I decided to take the messages bit by bit and see what I could make of them. I started with the information about the white blood cells.

I searched the internet for whatever it could tell me about what causes white blood cells to change their shape. The simple explanation was basic knowledge for any student of human biology. White blood cells are the cellular warriors in our microscopic internal defence force. They are the first responders which arrive at a site of infection or a wound to do battle with the potential invaders. Generally, in a healthy body, they travel in the blood throughout the body. If there is any potential danger: from infection via a wound which breaches the integrity of the system, from disease, from bacteria, from viruses, then the white blood cells rush to the site of the problem. They have to migrate through the walls of the tunnels, which are our blood vessels, to get to the danger zone. In order to do this they change their shape and literally squeeze through the interstitial spaces between the cells of the vessel walls.

The very fact that Katya, an elderly grandmother with no scientific education was able to impart remarkably specialised information about white blood cells amazed me. The very same information was overlooked by the scientists in Italy who were trained educated experts on cells. In their email to Channel 7, about which Katya knew absolutely nothing, they noted that the white blood cells did not look their normal shape. But they failed to see that the change of shape was due to them being suspended in the process of addressing inflammation. The New York Forensic Pathologist Dr Zugibe was fully aware of this. He had clearly concluded from his 2004 examination of the slide from the Buenos Aires case that the heart tissue was infiltrated with white blood cells similar to the cases where "a person has been severely beaten around the chest."

I was fascinated by what I was reading and needed to calmly, logically assess the data,

1. Somehow a study of why the white blood cells had changed their shape will lead to answering the questions I had asked about understanding the mystery of the mitochondrial DNA test results.

2. Jesus was also saying that these cases provide biological scientific data sets of the physical suffering of his torture and death.

3. The physical suffering which caused the inflammatory response in a person who died 2000 years ago is being seen today.

4. That same person who died 2000 years ago is alive today to explain the cellular results which led to his own death.

5. The cells, from the host and from the statue, are his because he himself explains them in terms of his lived experience as a man who was crucified.

My research into the role of white blood cells addressing trauma and injury led to some fascinating contemporary scientific reports describing additional effects of severe trauma and injury. Fragments of mitochondrial DNA are released from the damaged organs into the blood stream.[1] This circulating cell-free mitochondrial DNA is an additional marker of the body's immune response. What's more interesting is that this mitochondrial DNA can vary depending on the source of the injury. So cell free mitochondrial DNA which arises as a result of a heart injury will be different from mitochondrial DNA which arises from injury to lungs, or skin, or kidneys, for example. This mixture of DNA in a single individual is known as heteroplasmy.[2] And 'heteroplasmy' was the very term used by the Italian Laboratory scientists to explain the mixed mitochondrial data that had shown up in their tests.

My research became more interesting. I discovered that injury or trauma to the heart resulted in disproportionately large amounts of cell-free mitochondrial DNA being released, far more than in the case of injury to other internal organs and that this mitochondrial DNA routinely had the appearance of bacterial DNA.[3] I wondered if this could possibly be the explanation for the mixed DNA results that had shown up in the Italian lab tests on the Buenos Aires host-to-heart case. The Cochabamba statue tests had shown traumatized tissue intermingled with the blood. Could this be the explanation for mixed mitochondrial DNA data reported by the Italian lab? And could the difference between the cell free mitchiondrial DNA that is released from an injured heart and the cell free mitochondrial DNA released from traumatized tissue explain why the Italian lab failed to get a perfect match?

The issue seemed more complicated than the usual simple search to find an identity. Traces of specific suffering from specific injuries could well be incorporated into the stories these very unusual cells were telling us.

(1) The source of cell-free mitochondrial DNA in trauma
Kabilan Thurairajah, Gabrielle Daisy Briggs, and Zsolt Janos Balogh
https://www.ncbi.nlm.nih.gov/pmc/articles/PMC6002458/

(2) Very Short Mitochondrial DNA Fragments and Heteroplasmy in Human Plasma
Ruoyu Zhang, Kiichi Nakahira, Xiaoxian Guo, Augustine M.K. Choi, and Zhenglong Gu
https://www.ncbi.nlm.nih.gov/pmc/articles/PMC5095883/

(3) Increased circulating mitochondrial DNA after myocardial infarction.
Marte Bliksøen1, Lars Henrik Mariero1, Ingrid Kristine Ohm1, Fred Haugen, Arne Yndestad, Svein Solheim, Ingebjørg Seljeflot, Trine Ranheim, Geir Øystein Andersen, Pål Aukrust, Guro Valen2, Leif Erik Vinge2. Int J Cardiol. 2012 Jun 28;158(1):132-4
https://www.duo.uio.no/bitstream/handle/10852/48247/1/PhD-Mariero-DUO.pdf

CHAPTER 30

WHAT DOES IT MEAN

I had laid out in front of me the Menarini mitochondrial DNA test results on the Buenos Aires Host case (L1230) and the Cochabamba Statue case (L 1233). But I might as well have been looking at Godel's Incompleteness Theorem without having finished high school mathematics. All I could see was pages of numbers. I had no idea what to make of them.

It was late at night when I did what I had done so often in what is now my life's work. I decided that someone somewhere knows what this means. What's stopping me from knowing? Surely all it takes is persistent reading and learning until I know what they know about genetic data interpretation. Perhaps, I thought, in a moment of wishful thinking, along the way I might find my way to another professional that could help me. It would be most welcome because by now I knew that the advances in science regarding genetics were coming fast and furious. One day's certitude and consensus could be enhanced by tomorrow's breakthrough discovery.

Having regard to the testing approach taken by the Italian Lab and the amount of information gained from the testing, I reasoned that there was possibly a lot more to the results besides the fact that mitochondrial DNA might have been released by the organs of a person who had suffered trauma or injury.

The introductory remarks of the Italian Lab report went like this:

"mt DNA was analysed using next generation sequencing where only forensically relevant sequences are targeted.

Precision-ID mtDNA Control Region panel Kit (thernmoFisher Scientifiuc) was used as:

1. It is compatible with degraded samples

2. It covers whole Control Region (D –Loop), most informative value in forensic investigations

3. It targets positions 15,954-610 including the Hyper –Variable (HV) regions 1,11, and 111.

I soon learned that the Control Region of the genome contains specific information on the direct female line of a family and is used to help identify a person. Because Mitochondrial DNA mutates at a very slow rate, it is good for identifying genetic relationships over many years and many generations. The testing targets certain nucleotide positions within the Control Region and each position has a number assigned to it.

The testing results in this case, as for any sample tested, display those nucleotide positions relevant to the person. Not all the positions on the person's entire genome are presented but only those which vary from the nucleotide positions of a benchmark case. This benchmark was set as the standard by the genetic forensics industry and is known as the Cambridge Reference Sequence (CRS). The Cambridge Reference Sequence is the mitochondrial sequence of the first individual to have their mitochondrial DNA sequenced. The differences between the sample and the CRS are considered as mutations for the purposes of assigning a haplogroup to the sample. So what is a haplogroup?

A person's DNA is analysed and the results are then interpreted. One part of this interpretation involves placing an individual's DNA into a certain category. That category includes all the people who have the same genetic information at the same position on their DNA. They are distinct objective similarities with other people. They are genetic groups in the human population. These categories are called haplogroups.

They can be assessed by testing the DNA which reveals male ancestry or the mitochondrial DNA which records female ancestry. Y-DNA is passed on along the father's line, from father to son only, while mitochondrial DNA is passed on matrilinealy. But in this case it is passed on by a mother to all her children *regardless of whether they are male or female*. A grandmother, her daughter and her grandchildren, both male and female grandchildren, will all have the same mitochondrial DNA and be in the same haplogroup.

Each haplogroup shows us where we fit into the genetic family tree of all humans and groups us into ancient family "clans".

Because Mitochondrial DNA is inherited directly from your mother, and doesn't change in the process of sexual reproduction, the same maternal haplogroup will be shared with any relative that is in a direct maternal line with your mother. So, you, your mother, your mother's sisters, their children, your brother, your sister, and your maternal grandmother would have the same maternal haplogroup as you.

The global mtDNA tree has about 26 main branches which are classified by letters A-Z and each mtDNA haplogroup has further sub categories which are represented by letters and numbers. This amounts to *more than 4,000 different haplogroups* (van Oven and Kayser, 2009)

Once a person's haplogroup has been identified it can lead a surprising amount of information about that person's deep ancestry. The geographic regions from where the ancestors may have come and even the migration routes of those the ancient ancestors can all be inferred.

With all this in mind I studied the Menarini Lab report which presented details of genetic data of the Buenos Aires host case and the Cochabamba statue case. They were laid out in a spreadsheet. Focusing on the genetic information from the white blood cells from the Buenos Aires case I noticed that about 29 nucleotide positions had been listed in ascending order from 73 G to 16519C.

I then spent some time researching how one might be able to determine the haplotype from that genetic information. It turned out they were a perfect match for the essential defining markers within HVR1 and HVR2 of the Control Region for a known specific haplogroup. I did the same with the statue case. All the essential sequences to qualify for the exact same haplogroup came up. It too was in the very same haplogroup. I realized that the parallel of genetic markers must have been what prompted both Dr. Baggiani and Dr. Fontana from the Italian lab to comment "the samples from both cases seemed somewhat related."

Obviously what had prevented them from saying there was a perfect match in the results from both cases, was the additional data that had shown up in the results which I have attributed to cell free mitochondrial DNA released by organs of the body that have suffered traumatic injury which differed in the two cases.

The person whose heart muscle was found in the communion host from Buenos Aires, Argentina in 1996 had the same haplogroup as the person whose blood had shed from the statue of Christ which bled in Cochabamba, Bolivia in 1995. They are maternally related.

Out of more than 4000 possible haplogroups they both ended up in the same one, the one that describes a very specific group of people coming from the Middle East. I had for nearly 20 years been attempting to solve an identity riddle through DNA testing and it seemed now, finally, I had.

CHAPTER 31

GENEALOGY OF JESUS

> *In the language of Sacred Scripture 'heart' signifies the conscience that has become acutely perceptive and sensitive by virtue of love. It is a completely new way of knowing in union with me and with the community of my Apostles, a new experience of salvific solidarity.*
>
> Jesus dictates to Katya Rivias, *Crusade of Mercy*

Extracts from St Luke's Gospel given to me via Katya were an invitation to see what I could find. I began investigating what they could reveal.

The actual citation was from the second chapter of Luke. It was given in Aramaic, which was strange. Aramaic is the ancient language of Jesus and the Jews. If he wanted to speak to his apostles and friends he would have spoken Aramaic. But the only people present when the phone message appeared spoke English and/or Spanish. Was I meant to understand that Jesus was addressing the Jews? Or was this meant for the small remaining enclaves of people in Syria and Iraq who still spoke Aramaic? Was there something in the scientific findings which would especially be relevant to them?

I found the pertinent section of Luke and underlined the particular lines which came up on her phone:

And it came to pass, that in those days there went out a decree from Caesar Augustus that the whole world should be enrolled. This enrolling was first made by Cyrianus the governor of Syria. <u>And</u>

Joseph also went up from Galilee out of the city of Nazareth into Judea to the city of David called Bethlehem, because he was of the house and family of David, to be enrolled with Mary, his espoused wife who was with child. And it came to pass, that when they were there her days were accomplished that she should be delivered. And she brought forth her firstborn son and wrapped him up in swaddling clothes, and laid him in a manger because there was no room for them in the inn. And there were in the same country shepherds watching and keeping the night watches over their flock. And behold an angel of the Lord stood by them and the brightness of God shone around them and they feared with a great fear. And the angel said to them, Fear not, for behold I bring you tidings of great joy, that shall be to all the people. For this day is born to you a Savior who is Christ the Lord in the city of David. And this shall be a sign unto you. You shall find the infant wrapped up in swaddling clothes, and laid in a manger. And suddenly there was with the angel a multitude of the heavenly army praising God, and saying, Glory to God in the highest and on earth peace to men of goodwill.

It was as if a light went on. All of a sudden the mysterious citation from Luke made sense to me. The mitochondrial DNA results from two extraordinary events, a piece of bread turning into human heart tissue and a plaster statue bleeding, from different countries, from different times, both of which have an association with Jesus Christ, have the same maternal ancestry. Both were in Luke's words, *'of David's house and of David's family'*. I could see that it was loaded with meaning. I could see why Jesus would have wanted me to know about it. It seemed that the Italian lab results addressed the genealogy of Jesus, the Messiah, through his mother.

The segment of Luke's Gospel that was sent to Katya's phone was embedded in the following verses:

Now it happened that at this time Caesar Augustus issued a decree that a census should be made of the whole inhabited world. This census, the first, took place while Quirinius was governor of Syria, and everyone went to be registered, each to his own town.

<u>So Joseph set out from the town of Nazareth in Galilee for Judea, to David's town called Bethlehem, since he was of David's House and of David's lineage</u> in order to be registered together with Mary, his betrothed, who was with child. Now it happened that, while they were there, the time came for her to have her child.'

Biblical historians explain the census as an instrument with which the Roman Empire could efficiently levy taxes on its Palestinian Protectorate. Ancestry established Judaic legal rights to property and to inheritance and were therefore meticulously, transparently and jealously recorded. They were available for public scrutiny

Bethlehem, literally in Hebrew, 'the House of Bread', was called the City of David because that's where the great Jewish king came from. All the genealogical records relating to his lineage were held there. That is why when Joseph and Mary went to be registered for the census they went to Bethlehem. They were both of the house and line of King David.

Joseph and Mary were devout Jews. They knew very well that their particular ancestral line, the line of David, was prophesied to be the *only* one from which to expect the promised Messiah. And Bethlehem was where he was expected to be born. Seven hundred years before Christ, the prophet Micah declared, *'But you, Bethlehem Ephrathah, though you are small among the clans of Judah, out of you will come for me one who will be ruler over Israel, whose origins are from of old, from ancient times.'*

Two of the Gospel writers, Matthew and Luke include a genealogy of Jesus. Presumably they would have had access to those records in Bethlehem in order to write their accounts.

Matthew, the tax collector by profession, would have been very familiar with Jewish tribal genealogies since they were fundamental to assessing who owned what land and therefore who could be taxed and for how much. Matthew shows the ancestral line of Jesus through Joseph, his adoptive father, his *legal* father according to Jewish law. Luke on the other hand provides the *biological* line. He traces Mary's ancestral bloodline. Both genealogies go back through King David. Matthew was writing his account specifically for the Jews to convince them that Jesus was the expected Messiah from the line of David. Abraham, by virtue of a covenant with God, was the father of the Jews. For Matthew's case to be persuasive, that Jesus was indeed the long awaited Messiah, he has to first of all, before all other considerations, clear the first hurdle. Was Jesus of the line of David? So Matthew starts with Abraham and then goes forward in linear time to David and then to Jesus. Here is Matthew's version:

ST MATTHEW

CHAPTER 11-17

The book of the generation of Jesus Christ, the son of David, the son of Abraham: Abraham begot Isaac. And Isaac begot Jacob. And Jacob begot Judas and his brethren. And Judas begot Phares and Zara of Thamar. And Phares begot Esron. And Esron begot Aram. And Aram begot Aminadab. And Aminadab begot Naasson. And Naasson begot Salmon. And Salmon begot Booz of Rahab. And Booz begot Obed of Ruth. And Obed begot Jesse.

And Jesse begot David the king. And David the king begot Solomon, of her that had been the wife of Urias. And Solomon begot Roboam. And Roboam begot Abia. And Abia begot Asa. And Asa begot Josaphat. And Josaphat begot Joram. And Joram begot Ozias. And Ozias begot Joatham. And Joatham begot Achaz. And Achaz begot Ezechias. And Ezechias begot Manasses. And Manasses begot Amon. And Amon begot Josias.

And Josias begot Jechonias and his brethren in the transmigration of Babylon. And after the transmigration of Babylon, Jechonias begot Salathiel. And Salathiel begot Zorobabel. And Zorobabel begot Abiud. And Abiud begot Eliacim. And Eliacim begot Azor. And Azor begot Sadoc. And Sadoc begot Achim. And Achim begot Eliud. And Eliud begot Eleazar. And Eleazar begot Mathan. And Mathan begot Jacob.

And Jacob begot Joseph the husband of Mary, of whom was born Jesus, who is called Christ. So all the generations, from Abraham to David, are fourteen generations. And from David to the transmigration of Babylon, are fourteen generations: and from the transmigration of Babylon to Christ are fourteen generations.

Luke however was a universalist, writing for all mankind. His genealogy, the maternal line, starts with Jesus, goes back in time, way past David, past Abraham and even past Adam. He goes all the way back to God.

Only one human person was present at the annunciation by an angel of the birth of a Messiah. Only one person knew it would be a virgin birth. A scene so extraordinary, so unprecedented, so intimate and frankly so unimaginable, was narrated to Luke that its true author must have been actually present: the chief protagonist, the mother of Jesus herself.

Luke, by all accounts, a highly educated accomplished writer and thinker had a very unique problem: How does one record a classical traditional Jewish genealogy which always designates the father's line for a person who has no father? Women are generally not mentioned. No other genealogist had ever had to deal with the reality of a virgin birth before.

Luke was very inventive. He kept to the all-male pattern. He started off naming Joseph. But he qualified the name. He described Joseph as *'supposedly the father'*. This was another way of saying Joseph was not the father. Then he does a switch which would have been obvious to his contemporaries. He then *names Mary's father*, not Joseph's father. And from then on he's tracking the male ancestors of Mary. Here is Luke's version:

ST LUKE

CHAPTER 3 23-38

And Jesus himself was beginning about the age of thirty years; being (as it was supposed) the son of Joseph, who was of Heli, who was of Mathat, Who was of Levi, who was of Melchi, who was of Janne, who was of Joseph, Who was of Mathathias, who was of Amos, who was of Nahum, who was of Hesli, who was of Nagge,

Who was of Mahath, who was of Mathathias, who was of Semei, who was of Joseph, who was of Juda, Who was of Joanna, who was of Reza, who was of Zorobabel, who was of Salathiel, who was of Neri, Who was of Melchi, who was of Addi, who was of Cosan, who was of Helmadan, who was of Her, Who was of Jesus, who was of Eliezer, who was of Jorim, who was of Mathat, who was of Levi, Who was of Simeon, who was of Judas, who was of Joseph, who was of Jona, who was of Eliakim,

Who was of Melea, who was of Menna, who was of Mathatha, who was of Nathan, who was of David, Who was of Jesse, who was of Obed, who was of Booz, who was of Salmon, who was of Naasson, Who was of Aminadab, who was of Aram, who was of Esron, who was of Phares, who was of Judas, Who was of Jacob, who was of Isaac, who was of Abraham, who was of Thare, who was of Nachor, Who was of Sarug, who was of Ragau, who was of Phaleg, who was of Heber, who was of Sale,

Who was of Cainan, who was of Arphaxad, who was of Sem, who was of Noe, who was of Lamech, Who was of Mathusale, who was of Henoch, who was of Jared, who was of Malaleel, who was of Cainan, Who was of Henos, who was of Seth, who was of Adam, who was of God.

As a lawyer who had dealt with property law all my working life I found this fascinating and started to investigate. I learned that any ancient Jew's claim on land was based on the original tribal allocation of the land of Israel by God. To which of the original twelve tribes of Israel did one trace one's ancestry? Family history was intimately bound up in a claim to inheritance. Should someone assert that they had a right to a property, or to servants, or to an estate, or to crops, or to material possessions, that claim would only be considered licit depending on the claimant's ancestry. In order to transfer property, to sell property, to pass property on, you had to prove that you had the right to effect the transaction. There had to be verified ancestral bonds in order to conduct these transactions.

Certain tribes not only had a claim to certain land but also to certain jobs. Only those, for instance, of the tribe of Levi could fill the role of priests. And, most significantly, in the theocracy that was established by God in the land of Israel, only those people descended from King David could be eligible candidates for the long awaited title of Messiah. Being eligible naturally didn't make you a Messiah, but without it no other credentials mattered. It was the very first box to tick.

Genealogies were critical. It makes sense that it was customary, in a tradition of oral history, to memorise and recite entire lists of male ancestors. But they were also very accurately written down and archived from ancient times. Even throughout the Babylonian captivity and exile they were maintained in absentia from their homeland. The ancient historian Josephus writes that even Jews expelled and scattered throughout the world during the diaspora would continue to send the records of their children's births back to the towns where the family records were kept. They wanted an official genealogical record to be conscientiously maintained. And it was. That is, until 70 AD when they were all destroyed by the Roman General Titus Vespasian after the siege of Jerusalem.

Jesus had a mother and a foster father, Joseph, his legal father. He had two grandfathers. Through his mother he had a grandfather named Heli, (also known as Joachim) and through his foster father he had a grandfather named Jacob. The names of Jesus' descendents are the same on both maternal and paternal sides, from Abraham, the father of the Jews, all the way through many centuries until

David. Then the genealogies diverge. One branch comes through David's son Solomon. The other comes through David's other less famous son, Nathan. The Solomon branch leads to Joseph. The Nathan branch leads to Mary.

If the Jews decided to insist on the Messiah having a Davidic bloodline as well as a Davidic legal genealogy, then the two genealogies together provide all the detailed irrefutable credentials needed. These genealogical credentials were so impeccably watertight that of all the many reasons invented to discredit him the enemies of Jesus never once, not during his public ministry, not during his trial, nor after his death in the decades that followed, even obliquely attempted to smear him with the one accusation that would have eliminated him immediately from his claim to being the Messiah: that he was not of the Davidic line. The enemies of Jesus knew that the genealogical record was very public, very valuable, and very old. That King David was his ancestor was rock solid.

Not one of his powerful enemies protested when the multitudes in Jerusalem cried out in acclamation when a poor man riding a donkey entered the city gates for the great religious festival on Palm Sunday,

'*Hosanna to the <u>Son of David</u>. Blessed is he that cometh in the name of the Lord. Hosanna in the highest*'
Matthew (21:9)

The stakes were high for them. A long awaited Messiah was the culmination of Israel's history and the fulfilment of God's promise of redemption. Their hope, their purpose and their eternal destiny were inseparably connected to him. They still are.

CHAPTER 32

THE WAY I SEE IT

My Heart was a small piece of the Immaculate Heart of my mother. My blood was her blood. My wounded flesh issued from her flesh.

Jesus dictates to Katya Rivas, Crusade of Mercy

I had asked Jesus through Katya for help in understanding the results of the Italian lab. It was a question that had to do with ancestral maternal DNA testing and him. Jesus knew that I had been working for nearly twenty five years trying to get to this critical point in the investigation of these cases. He knew its importance. On previous occasions Katya conveyed directions and inspirations in this ongoing investigation. In response to my question this time God led me to the Gospel account of the ancestry of his mother, the Virgin Mary, and her blood line back to King David and beyond.

Curiously the phone message from Jesus was written in Aramaic, a language which Luke may have spoken, but which he, a Gentile Christian, being neither Jew nor of Hebraic lineage, certainly didn't write his two major works in. Both his Gospel and the Acts of the Apostles were written in Greek. What's more it was a refined, literary form of Greek which underscores the view that Luke was well read and highly educated. According to St Jerome's commentaries they were written while Luke was living in Achaia, in what is now Greece, to a Greek man called Theophilus, ('theo' and 'philos' are Greek words). By a large consensus historians and biblical scholars agree that Luke wrote in Greek, not least of all because the most ancient manuscripts are in Greek.

Why then was Jesus quoting Luke's gospel in Aramaic? And if Jesus was addressing the Jews why employ an ancient language instead of Hebrew?

I was becoming very curious. Was God telling me that the ancestral clan haplogroup we had discovered in our testing is somehow related to the Virgin Mary and her ancestral blood line back to King David? And if the haplogroup is that of the ancestral line of the Virgin Mary, who will we find back in history to have been the original maternal source of that haplogroup?

Luke traces the paternal blood line of Mary back to Adam. Will Mary's maternal haplogroup be that of the real Eve, wife of Adam? Was Jesus hinting at the pursuit of further ancestral DNA studies to confirm that He, through Mary, does have a direct ancestral blood line back to King David.

These questions remain open to be answered.

The Old Testament has references to the coming of a Messiah and that he would be a descendent of David. The Jews, even today, do not accept Jesus as the Messiah who was prophesied to come because, they argue, his link to David comes from his adoptive father Joseph, and not from a blood line. This is the primary disqualifier. But if Jesus does have maternal royal Davidic bloodline and that this can be demonstrated scientifically by genetic testing, then a whole new chapter in Christian history commences.

I pondered the second sentence in the message on Katya's phone: "There were some shepherds in the same region lodging there and keeping nightly watch over their flock…" I found it in context in Luke, Chapter 2:

Now there were shepherds in that region living in the fields and keeping the night watch over their flock. The angel of the Lord appeared to them and the glory of the Lord shone around them, and they were struck with great fear. The angel said to them, "Do not be afraid; for behold, I proclaim to you good news of great joy that will be for all the people. For today in the city of David a savior has been born for you who is Messiah and Lord. And this will be a sign for you: you will find an infant wrapped in swaddling clothes and lying in a manger." And suddenly there was a multitude of the heavenly host with the angel, praising God and saying: "Glory to God in the highest and on earth peace to those on whom his favour rests.

Jesus, I believed, must have drawn this passage to my attention for a reason; perhaps in support of his claim to be the Messiah. This angelic supernatural announcement of the arrival of the Messiah, King of the whole world, is given to the humblest of people, shepherds. They become His witnesses. Hadn't I, as a lawyer, been assiduously concerned throughout these decades with collating witness accounts? Some, like the shepherds, were uneducated and some, like the scientists and the priests, were highly educated. All had significant contributions to make as witnesses to the remarkable events and discoveries concerning the Eucharistic host and the statue which wept and bled.

The two separate phrases Jesus gave Katya, complement each other as evidentiary elements. The first draws attention to the records in Bethlehem which document Mary's ancestral line back to King David. The second tells of the birth of Mary's son, the Messiah announced by an Angel and verified by shepherd witnesses.

These are my interpretations, as a layman with no scientific or theological credentials, of what I believe the Italian lab results show when seen in the light of the passages from St Luke. My assessment is not as an expert but neither is it from ignorance. I have studied the science surrounding these cases and assiduously sought expert opinion. I have relied on statements made by DNA experts and pathologists who have examined these cases over some twenty five years. I am a lawyer not a scientist and what I have had to say will need to be vigorously examined by the appropriate experts as a next step in this process. I have examined the Italian Lab results in the light of what was given to Katya by Jesus during mystical experiences in response to questions I posed.

Combining both parts of the message on Katya's phone with what was achieved by the Menarini mitochondrial DNA tests, I am able draw this rough sketch of the person (or persons) who is identified in the material tested.

* The person or persons have no human genetic profile. This was established by Nuclear DNA testing of our samples numerous times.

* The person in both cases has suffered traumatic injuries. This has been revealed in forensic pathology testing and from mitochondrial DNA released by affected organs showing up in advanced DNA analysis with Next Generation sequencing technology.

* In both cases Mitochondrial DNA testing using single cell analysis with Next Generation Sequencing has yielded considerable data on the maternal ancestral line of the person in both cases to reveal a known haplogroup. That haplogroup is the same for both persons.

This brief sketch I have drawn relates potentially to Jesus Christ, who, according to the Catholic faith, is present in every consecrated communion host, who, was tortured, crucified, died, and rose from the dead two thousand years ago. After his resurrection his blood stained burial cloth was retrieved and venerated for two millennia. That cloth is known as the Shroud of Turin.

If the blood that has been examined in our two cases is truly the blood of Jesus Christ and if the blood on the Shroud of Turin is also that of Jesus Christ then the results from identical testing on both should match each other. No other person has ever not had a genetic code. Not having a genetic code is unique in human history. If they do match in this singular regard highly qualified scientists could be enjoined by the Catholic Church to interpret the DNA data anew and to evaluate my assessment of the implication of the results that I have presented in this chapter.

I am confident that there will be a match because of a remarkable co-incidence. When I was going through some old videotapes of interviews I came across one I had forgotten about. It was an interview with Katya in October 1996 in Miami. In it she told me about a message Jesus dictated to her a year earlier on the 14 April 1995. That date immediately caught my attention. It just so happened to have been the same day, 14 April 1995, that the very first sample of blood was taken from the statue, the same sample I still have in my possession. Jesus said to her, *"I want the blood that is wiped from my Image to be given to the Church authorities and for it to be compared with the blood of my Shroud. It is time for the lies to be buried and for the truth to be revealed."*

In the interview I asked Katya what she understood by that message?

"I understand", she replied, "that when the blood from the statue is compared with the blood on the Shroud of Turin they will find that the blood is from the same person and that this will assist in the authentification of the Shroud of Turin as the true burial cloth of Jesus Christ."

"Do you realize," I asked, "what the consequences for you will be if the blood does not match?"

She replied with a soft smile, "I trust in My Lord."

CHAPTER 33

SHROUD

*"It is time for the lies to be buried
and for the truth to be revealed"*

Relics are venerated remains. Veneration is distinct from, and not to be confused with, worship. God alone is worshipped. People who lived lives of heroic virtue, saints and martyrs, have traditionally been venerated. And so too have material artifacts associated with them. Various objects they used or wore may be venerated as relics of these spiritual champions. No artefact, however, is esteemed higher as a relic than their physical deceased bodies, once exemplary temples of the Holy Spirit. While the Catholic Church refrains from officially pronouncing that any particular relic is authentic, the honour given to relics, which are reasonably believed to be genuine, is encouraged. The holier the saint the more venerated the relic. None is more holy than Jesus. But is there on earth any remnant of his body? Could it be possible that something physical remains, a relic, of a man who died and was later seen alive and then ascending to heaven by over five hundred witnesses?

For two thousand years Catholics have venerated the bloody burial cloth of Jesus Christ as the most sacred of all relics. That cloth is known as the Holy Shroud of Turin. It is the most famous, most treasured relic, and also the most exceptional. Unsurprisingly it is also the most examined sacred object in history. On what is essentially a 14 foot long piece of linen is a full length life size image of a 5' 11" man covered in wounds. These wounds conform to scourging, crucifixion and having thin sharp objects puncture the skull. Also on it, clearly visible, are multiple stains, created, forensic scientists and chemists now know, by blood.

On the evening of 28 May 1898, when the new technology of photography was in its infancy, the shroud was photographed for the first time by Secondo Pia. At around midnight he took the photographic plate to his darkroom to develop it. What he saw on the plate so shocked him that he nearly dropped it. On the first ever negative plate of the Shroud, a positive three dimensional image of a man emerged. To the naked eye a real man's face could be clearly seen. The Shroud image had been, all along, a photographic negative, one which could only have been 'developed' into a lifelike representation once technology had advanced to the point where photography was invented. Science, in this case the science of optics and chemistry involved in the making of cameras and photographs, had resulted in a latent revelation which had lain hidden for centuries. There is no other image in existence with this property.

This astonishing photographic 'faculty' is not found in any other cloth, ancient or modern. It has been confirmed by other photographers since then. In the 1960's Leo Vala, a professional London photographer used slide projectors to beam negatives of the shroud onto malleable plasticine. The resulting three- dimensionality provoked him to exclaim, "I have been involved in the invention of many complicated visual processes and I can tell you that no-one could have faked that image. No-one could do it today with all the technology we have. It's a perfect negative. It has a photographic quality that is extremely precise."[1]

In the late seventies an electronics engineer, Peter Schumacher, developed a sophisticated topographical device for NASA's space program called the Interpretation Systems VP-8 Image Analyzer. It translated a monochromatic tonal scale of the photographed surfaces of planets into layers of vertical relief. When a normal photograph is analysed by the VP-8 it is invariably collapsed and distorted. But when physicist, Dr John Jackson analysed the shroud image in the system it was consistently and uniquely three-dimensional. Schumacher described the moment he first saw the shroud results from his system.

"A true three-dimensional image appeared on the monitor. The nose ramped in relief. The facial features were contoured properly. Body shapes of arms, legs and chest and the basic human form... I had never heard of the Shroud of Turin before that moment. I had no idea what I was looking at. However the results are unlike anything I have processed through the VP-8 Analyzer before or since. Only the Shroud of Turin has ever produced these results."[2]

He went on to comment on the ever growing and bewildering discoveries about the Shroud in the area of scientific imaging.

"One must consider how and why an artist would embed three-dimensional information in the 'grey' shading of an image when no means of viewing this property of the image would be available for at least a thousand years after this was done. One would have to ask why is this result not obtained in the analysis of other works?... Why would the artist make only one such work requiring such special skills and talent, and not pass the technique along to others? How could an artist produce this work, show these results before the device to show the results was even invented?"[3]

"No method, no style, and no artistic skills are known to exist that can produce images that will induce the same photogrammatic results as the shroud image induces... The shroud image exhibits some properties of photographic negatives, some properties of body frame imaging and some properties of three-dimensional gray scale encoding. It is 'none of these' and represents portions of 'all of these,' and more."[4]

In the 1980's a chemist at Michigan University, Dr Giles Carter added to the intrigue when he discovered that the Shroud displayed properties shared with long wave X-rays. A skeletal structure of bones and hands and two rows of teeth were uncovered and made visible.

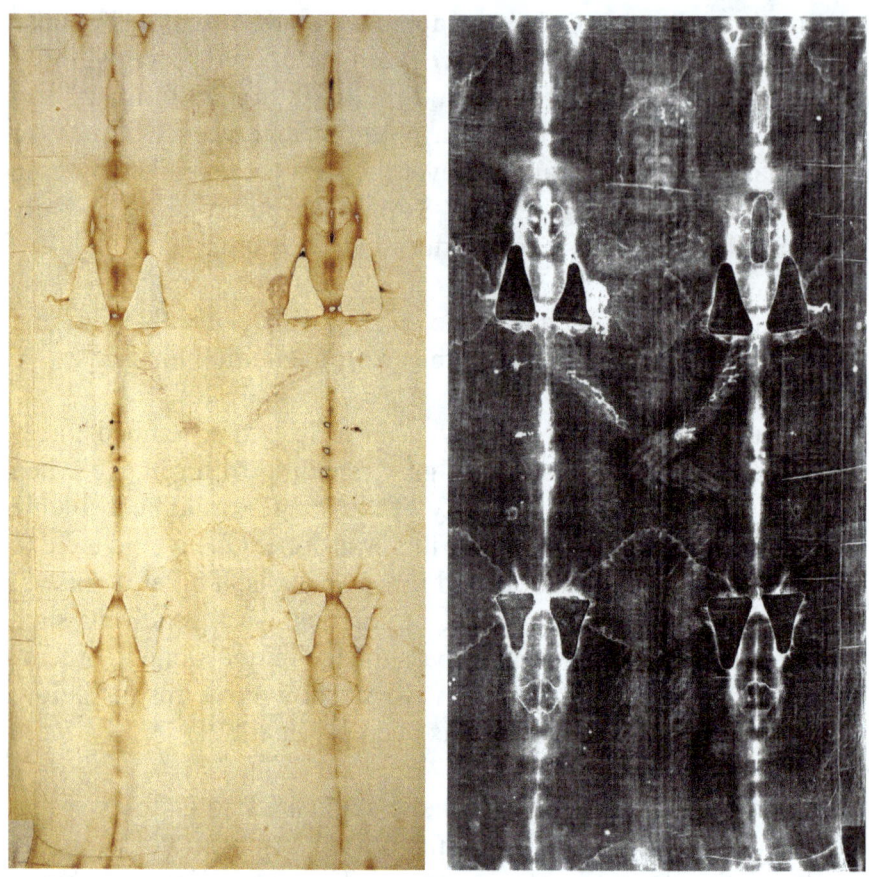

Front view and negative view of the shroud. ©1978 Barrie M. Schwortz Collection, STERA, Inc.

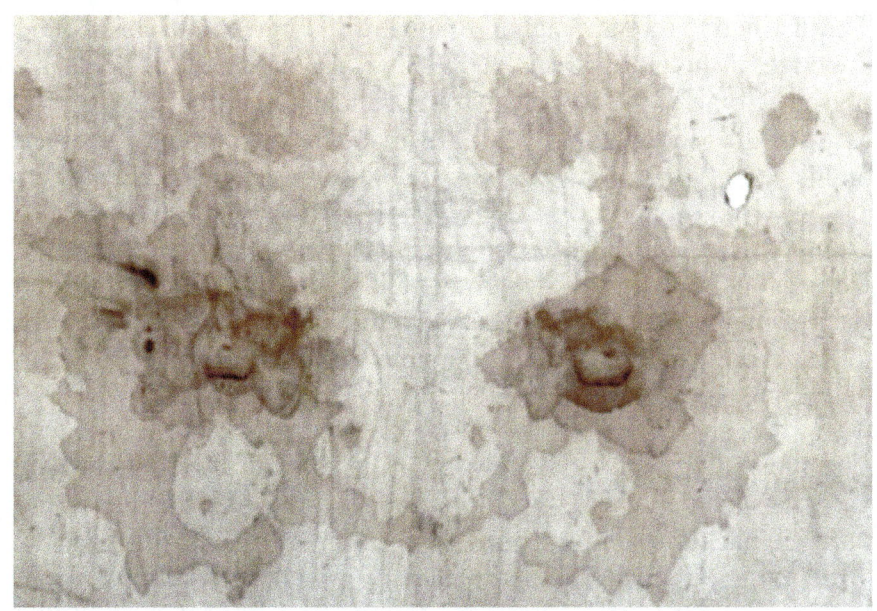

Blood Stains on the Sudarium (Courtesy Goya Productions, Madrid)

Dr Alan Adler at Conference on the Shroud (Nice, France, 1997)

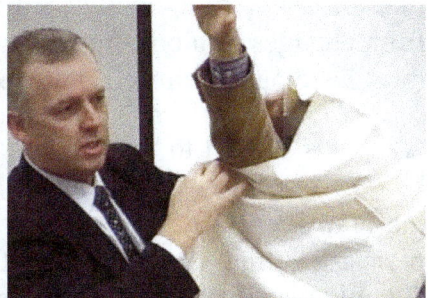

Mark Guscin demonstrates how the Sudarium was used on Jesus crucified

The image on the Shroud, though remarkable, is one thing. But it is not the only thing which is of interest to the scientific community. The blood is another. This is what science can tell us so far about the bloodstains:

Microchemical, fluorescence and spectrographic tests reveal that the stains are blood. They resulted from real bleeding from real wounds of a real body that came into direct contact with the linen.

The distinctive forensic signature of clotting, (red corpuscles around the edge of the clot and a clear yellowish halo of serum) is clearly present. Some of the blood flow was venous and some was arterial. Most of the blood flowed while the man was alive and it remained on his body. Some, however, clearly oozed from a dead body.

Pathologists who have studied the Shroud agree that the stains are from a man lying on his back on top of the fabric. It was then draped over his head, covering his face and entire length of the body, down to his feet. This back and front view have enabled scientists to determine the position and the nature of the man's wounds.

The stains along once-outstretched arms show that blood from the victim's wrists flowed down along the forearm, past the elbow and onto the back of the upper arm. The outline of bloodstains near the armpit suggests blood pooled before dripping to the ground. From the angles of the stains the forensic experts have determined that this blood flowed while the man was upright with his arms at angled like the hands of a clock at ten minutes before two. Also evident from changes in bloodstream angles is the likelihood that the man must have pulled himself up repeatedly.

A serious wound to the chest is very clear. And interestingly, the blood from this wound shows that the wound was inflicted post mortem. Present also are stains from a clear bodily fluid such as pericardial fluid or fluid from the pleural sac or pleural cavity. Cumulatively these findings indicate that the man was stabbed near his heart after he died.

None of the stains, the tissues, the body fluids, or the patterns of blood flow could have been manufactured by brushing, daubing or pouring human blood onto the shroud. The blood, rich in bilirubin which the body produces under extreme trauma, is unquestionably the blood of the man whose lifeless, crucified body was enshrouded in the cloth.

Dr Alan Adler, a research chemist from Western Connecticut State University, an expert on porphyrins, the types of coloured compounds seen in blood, and Dr. John Heller studied the blood flecks They converted the heme into its parent porphyrin, and they interpreted the spectra taken of blood spots. In addition, the x-ray-fluorescence spectra showed excess iron in blood areas, as expected for blood. Microchemical tests for proteins were positive in blood areas but not in any other parts of the Shroud.

At a conference in Nice on 11 May 1997 I filmed Dr Adler. He said, "There is no dye, pigment or stain making up the image of the man whose body you see on the cloth. It is not a painting. It simply represents chemically a dehydrative oxidative process of the cellulose of the fibres constructing the weave of the fabric of the cloth. This discolouration is only one fibre deep on the crowns of the threads comprising the weave, which I point out to you is half the thickness of a human hair. The blood images on this cloth were put there by this clot coming in contact with a man who was wounded and died a traumatic death. When you analyse the blood wounds you observe that you are not looking at whole blood, you are looking at exudates of clotted wounds and furthermore there are some very peculiar features in the chemistry of these wounds and their composition: they contain an abnormally large amount of bilirubin. There is a very simple explanation for that. You are looking at a man who died a severely traumatic death prior to the shedding of this blood. In other words he was in traumatic shock, the kind of thing you would get if you were beaten and crucified. Under these conditions you get an enormous so-called icric index of the bilirubin to the haemoglobin into the blood which, when it forms the clot, will be in the exudates from the clot. And will make the kinds of wounds you see, in particular the red colour."

The blood on the shroud is very old and almost certainly degraded. If new samples were taken for DNA testing, and if a genetic code was discernible, this in itself could not be said to be the genetic code of God. It would simply be the genetic identity of a crucified individual. As history teaches, many hundreds of men were crucified by the Romans in Palestine at the time, albeit not men who would be expected to have sharp thorn-like puncture wounds on their heads.

Dr Adler put the big question in a nutshell. "The hypothesis, 'Is this Jesus Christ?' is not experimentally testable. We do not have a laboratory test for Jesus Christness."

If the blood that has been examined in our two cases is truly the blood of Jesus Christ and if the blood on the Shroud of Turin is also that of Jesus Christ then the results from identical testing on both should match each other. No other person has ever not had a genetic code. Not having a genetic code is unique in human history. If they do match in this singular regard highly qualified scientists could be tasked by the Catholic Church to interpret the DNA data anew and to evaluate my assessment of the implication of the results that I have presented.

In spite of much of the work done on the Shroud, in spite of hundreds of books and scientific articles, in spite of most of these affirming the authenticity of the Shroud as the burial cloth of Jesus, there is still a general perception, promoted by the media, that the Shroud is a fake. And yet in the face of negative public opinion a strong body of evidence has emerged to support the authenticity of the shroud. It has come from a Spanish team of experts known as the Centro Español de Sindonología (the Spanish Centre for the Study of the Shroud). This team had initially been investigating an entirely different relic. It had carried out multi-disciplinary studies which cast a whole new light on the authenticity of the Shroud. The other, less well known relic is known as the Sudarium of Oviedo.

(1) Wilson, I., *The Shroud* (Bantam Press, Great Britain, 2010) p 21

(2) Ibid p 22

(3) Ibid p 22-23

(4) From transcript of filmed recording of Dr Alan Adler by Ron Tesoriero 11 May 1997

FaÓoÉ ʟ‰

Zugibe, F, *The Crucifixion of Jesus: A Forensic Enquiry* (M. Evans and Company, New York, 2005) 211-224

Rogers, R. N. *Frequently Asked Questions on the Shroud* (2004)
http://www.shroud.com/pdfs/rogers5faqs.pdf viewed 21 March 2013

Heller and Adler, Applied Optics 19, 1980
Wilson, I. and Schworz, B. The Turin Shroud (Michael O'Mara Books, London 2000)

CHAPTER 34

SUDARIUM

Is there no test for Jesus Christness? Is what Dr Adler proclaimed still the case? It may have been true at the time he said it but there are now other studies conducted by the Centro Espanol de Sindonologia on a second piece of cloth to take into consideration. There may indeed be tests, plural, for "Jesus Christness" thanks to this second cloth called the Sudarium of Oviedo.

The Sudarium Domini, the 'sweat cloth of the Lord', is in Oviedo in Spain, in a vault, the Camera Sancta of the Cathedral of San Salvador. It is another blood stained linen cloth, about the size of a towel, or napkin, the one purported to be the second cloth in the tomb of Jesus mentioned in the Gospel of St John 20: 6, *'then cometh Simon Peter, following him, and went into the sepulchre, and saw the linen cloth lying, And the napkin that had been about his head, not lying with the linen cloths but apart, wrapped up into one place.'*

This one, the Sudarium Domini, was found still rolled up in a corner when he, John, the only apostle who was a witness to both the crucifixion and burial of Jesus, entered the tomb on Sunday morning. It was still in the same place where it had been left. The shroud which had covered the body on the stone mortuary shelf is described as being 'flattened' or 'deflated' (from the Greek 'keistha') in contrast with the sudarium which was still rolled up and unchanged in shape and position from the time the disciples left later on Friday evening. The tomb was then sealed.

Whatever it was that John saw made him declare in the next verse, *'He saw, and believed. Until this moment they had not understood the scripture, that he must rise from the dead.'* John goes to surprising length to detail exactly what he saw: the contrasting position and condition of the burial cloths. And since the absence of a corpse in itself was not a confirmation of resurrection, John must have drawn this conclusion from, as he himself says, what he saw. Could it be that what he saw was a life-sized image of a man on an unusually flattened, blood stained cloth that had been covering his body when John left the tomb last? Could it be that by contrast the sudarium that had been rolled up and discarded apart from the body was still exactly where and how it had been left.

The Sudarium enjoys an historically straightforward well-attested provenance documenting its passage from Palestine to where it is now, and has been, for over a thousand years. For much of the Sudarium's history, keeping it safe has meant keeping it secret. In 1075 it was removed from under an altar where it was concealed when the Moors invaded Spain in 710. Then in 1113 it was placed in a silver chest on the orders of King Alfonso VI who, at the same time, included it in a detailed inventory of objects sacred to the Church.

Unlike the more dramatic shroud, the sudarium bears no image. It shows bloodstains of varying colour densities. In 1989 a team of scientists and experts agreed to cooperate in studying the sudarium under the auspices of the Centro Espagnol de Sindonologica, each working in his area of specialization. The results have been very interesting indeed.

Studies have shown that a sudarium was used in compliance with ancient Jewish funereal rites which accord with very deeply held religious beliefs concerning the soul and blood. Jewish custom insisted on burial. To remain unburied was a solemn dishonour. Since blood was considered to contain the life of the soul it was revered as belonging to God. Blood constituted the material component of the life of a soul created in the image of God. Atonement could be made by blood alone, and the sins of person who had died were expiated through the decay of the blood after death. This meant that in the cases where loss of blood was a factor in the death of a person, that body would not be ritually washed. Keeping as much blood was like keeping money to pay for the sins of the soul. In this way as much blood as possible was retained as an expiatory payment for sins in the hope of final atonement. Even more concerted attempts were made to preserve whatever blood was remaining by anointing the unwashed body with aloes and myrrh, an aromatic and a resin.

Jewish tradition also insisted that a face disfigured at the time of death be covered. And what the Sudarium is claimed to be: the cloth that was wrapped around the head of a crucified man before he was lowered from his cross.

The Spanish forensic team has established that it covered the head of a male who was already dead; one who died in an upright position with his head tilted 20% to the right, 70% forward and with outstretched arms, a position which was maintained for 45 minutes after death. The victim died of asphyxiation as shown by the 5:1 ration of pulmonary edema on the cloth. Stains show that a hand clasped the cloth over the nose of the victim to staunch the flow of post mortem edema which surges through the nostrils when a body is moved shortly after death. This action conforms to the value placed on blood as the currency of atonement by devout Jews.

Analysis of vital blood and postmortem blood – blood which flowed while the victim was alive and blood which seeped from the corpse – bear uncanny concordance with the time frame of events outlined in the gospel accounts. Two hours elapsed from the time of death to burial. The cloth was folded to cover the head of a man whose head lay angled towards his chest making the arm an obstruction to it being knotted: it was removed and re-wrapped after the body was no longer upright and the obstruction no longer a problem; that it was laid horizontally for 45 minutes before being finally removed. [1]

Even more interesting are the results obtained by Dr Alan Whanger when applying a polarised image overlay technique on both the shroud and the sudarium. Comparisons of the type and positions of blood stains, and comparisons of the facial features reveal over 70 points of coincidence on the front and 50 points on the back. This represents an extremely high rate of alignment known as consilience. The nose is exactly the same length, swollen and displaced to the exact degree in both cloths. Puncture wounds from sharp objects on the head coincide. Blood and lymph densities coincide. There was a remarkable coincidence of locations of blood stains on both cloths and that this was evidence that the two cloths had come into contact with blood oozing from the injuries of the same crucified man.

The studies on the Sudarium of Oviedo are monitored by Alfonso Sánchez Hermosilla, a researcher in UCAM, the Catholic University of Saint Anthony of Murcia, the coroner from the Legal Medicine Institute of Murcia, headmaster of the Sindonology Research Spanish Centre (EDICES) and an adviser at the International Sindonology Centre in Turin. He wrote about the most recent investigations of the Sudarium using optic and electronic microscopy.

'It shows the presence of structures compatible with highly damaged red blood human cells, some of them haemolyzed, and also with fibrin clots free from hematic structures.'

The forensic medical study describes in detail the tissues and organs the piercing object ran through, in its hypothetical trajectory. It adds that 'particularly the heart right auricle, in cadavers of people that have suffered a long agony, very often present big blood clots, very similar to those that formed the side stain in the Shroud of Turin' and that, ' in the piercing of the right lung, the weapon opened its way through the intra-parenchymal airways, and as a consequence of this, some of the organic fluids made its way in a rising direction, due to the intra-thoracic pressure, caused by the kinetic energy the weapon transmitted to the cadaver. These fluids also travelled through the cadaver's mouth and nose and caused new stains in these areas of the Shroud of Oviedo. Of course, when the weapon was removed, these fluids went out through the entry-exit holes.'

The Centro reported that, 'These discoveries are compatible with an intense physical maltreatment, with multiple traumas that produce bruised wounds, bleeding wounds, sharp wounds and bruised wounds, what probably includes flagellation in the Roman manner using a flagrum taxilatum. The discoveries that have been found open new areas of research that were unexpected until now. A priori they seem to be really promising and include new stains that were unknown until now. For this reason, it is reasonable to believe that it would be opportune to carry out new direct research in the future on both Relics and to relate the discoveries that have been verified in the Sudarium of Oviedo to possible matches to the Shroud of Turin.'[2]

Anthropologically, and according to microscopic forensic analysis both cloths covered the same victim. If the sudarium is a fake then a medieval fraudster would have had to fabricate both the Shroud and the Sudarium simultaneously. He would have had to use an identical human model for both the Shroud and the Sudarium, that is, the same corpse, with the same forensic footprint, who died at the same time by the same means, Roman crucifixion, and who had very unusual multiple sharp puncture wounds on his head totally alien to the standard method of Roman crucifixion.

I met Dr Sanchez Hermosilla at a Conference on the Shroud of Turin in Valencia in April 2012 where he delivered a paper on his forensic analysis of the two cloths. By 2019, after I had mitochondrial DNA results from the Italian lab, it was time for the investigation to get even bigger. I considered that it was the right time for the latest genetic findings from our cases of Buenos Aires and Cochabamba to be compared with similar tests on the blood on the Shroud. It was time for another meeting with Dr Sanchez Hermosilla.

Jorge Manuel Rodriguez, the president of the Spanish Centre for Studies on the Shroud in Valencia organised for us to meet.

---

(1) For a detailed statement of findings of the Spanish Forensic team, see Publication El Sudario de Oviedo, (2000 Ediciones Universidad de Navarra, SA EUNSA) by Jorge-Manuel Rodriguez Almenar.

(2) https://www.shs-conferences.org/articles/shsconf/pdf/2015/02/shsconf_atsi2014_00007.pdf

CHAPTER 35

MEETING IN MURCIA

On 17 September 2019, accompanied by Dr Colin Summerhays, a surgeon, we met Dr Sanchez Hermosilla at his favourite little restaurant in the town of Murcia. He was taken aback when I told him that all I wanted was an hour of his time and for that hour I had flown across the world from Australia for 24 hours and taken a four hour train ride south of Valencia. That was how important it was to me. I discussed the work I had been doing and the results of the testing of our samples from Buenos Aires and Cochabamba. I explained how both pathology and DNA testing on samples from both cases revealed that the person in question had suffered traumatic injuries, possibly from having been severely beaten. These were similar conclusions that he, as a forensic medicine expert, had arrived at in his analysis of the blood on the Sudarium and the Shroud of Turin.

He was particularly impressed by the results of mitochondrial DNA testing we had done on the control region of the human genome which revealed valuable data on the maternal hereditary line. The value of conducting identical testing on the Shroud was immediately apparent to him. The results we had could provide a base line for comparison. I was surprised that he was not aware of the latest technology we had used with the Italian lab which could do single cell analysis. This confirmed again how very new the technology was.

17 September 2019 I show the tests results of the Buenos Aires and Cochabamba cases to Forensic Pathologist Dr Sanchez Hermosilla in Spain.

Up until now the biggest nightmare DNA researchers always faced was the possibility of contamination. For shroud researchers the nightmare was amplified by the source being touched by adoring hands and kissed by the faithful for over 2000 years already. Single cell analysis might, Dr Sanchez agreed, finally provide a breakthrough in being able to get reliable results from future DNA testing of the Shroud.

It is now time for the Shroud and the Sudarium to be tested by the same Italian laboratory using the same tests. Single cell extraction and analysis is minimally invasive. No more cutting up of either cloth is required. Permission for scientific examination and testing of the Shroud rests with the very same man who, then as Archbishop Bergoglio of Buenos Aires, commissioned the investigation into the transformation of a consecrated host in his very own diocese twenty five years ago. And since the testing would include genetic analysis of uncontaminated single cells the entire project could fall into the eminently scientific hands of Dr Francis Collins. Dr Collins is an esteemed member of the Pontifical Academy of Science and no stranger to the Vatican. His authority on the subject of genetics is undisputed since being honoured as the first geneticist to map the entire human genome.

Because the blood is so old it would not be surprising if DNA analyses of both the shroud and the sudarium were unsuccessful in eliciting a nucleic genetic profile. I would expect this to be the case for the very reason that every time I have had nucleic DNA tests done they have come back with spectacularly unique inexplicable results showing no genetic code whatsoever.

But science has available to it another source of DNA in the cell, one which is likely to be of particular interest in analysing the Shroud and the Sudarium: mitochondrial DNA. This source is more robust than the nucleus and has proven far more suitable for the study of ancient relics. It has also been of very particular interest in analysing the cells found in the host from Argentina and the statue from Bolivia. Both tests reveal similar data relating to ancestry. Both tests indicate physical suffering.

All the scientific evidence accumulated across the samples submitted for testing from the host-to-heart and blood case, as well as from the statue-to-skin and blood case should be considered alongside results from future tests on the Shroud. Asking whether there is a test for 'Jesus Christness' may not be quite the right question to be asking. Instead we should ask are there *tests* – plural – which contribute by force of their combined evidence to a verdict?

When I asked him if he had any plans to get access to the Shroud to take samples for further testing he became noticeably subdued. He answered in a tone of defeat.

Dr Sanchez Hermosilla: "We have tried a number of times to get permission from the Church authorities to take samples for more testing and we have been refused. Ultimately, it is Pope Francis who has to give permission and he seems reluctant to do so. I think there has been so much controversy about the Shroud in recent times and I sense that there is a reluctance on the part of the Church to get involved and fuel more controversy. We are very constrained in how we can deal with some of the unfair criticism about the

Shroud without doing more testing with new samples. The type of DNA testing you are suggesting is certainly something that should now be done. I don't think it will be easy to get the consent of Pope Francis so we really are at a standstill until that happens."

Me: "Well maybe I have an approach that is worth a try. It is somewhat coincidental that the only person in the world who can give consent to this next step in the study of the Shroud, is the same person who wanted an investigation into the Eucharistic Miracle of Buenos Aires that I am involved in. Perhaps if I wrote to him asking for consent as the final part of the investigation of the Buenos Aires case, the one he himself asked for, he might relent and grant permission. It seems reasonable that if we have got genetic information of the person whose heart tissue was found in the transformed host, that information should be compared with information that might come from doing the same tests on the Shroud. The genetic information must match if it comes from the same person, if in faith we believe it is Jesus Christ."

Dr Sanchez Hermosilla: "For you to approach Pope Francis on that basis is a great idea."

Me "Well that is what I will do."

Before I said goodbye to Dr Sanchez I told him briefly about Katya Rivas and her part in my story. I started by describing the strange coincidence which occurred on the day the very first sample was taken for testing. It was 14 April 1995 and the very first scrapings of what looked like blood were taken from a 'bleeding' statue in Bolivia. Katya had no knowledge of the taking of the sample. She was miles away. On that very same day she received the following message from Jesus:

"I want the blood that is wiped from my image to be given to the Church authorities and for it to be compared with the blood of my Shroud. It is time for the lies to be buried and the truth to be revealed."

I had consciously restrained myself from telling Dr Hermosilla-Sanchez this until the end of the meeting, and even then I was hesitant, because a message from Jesus was hardly something I would raise at a scientific meeting with an esteemed forensic scientist. I did not know how he would take it. I told him that I had relayed the story in my book *Reason to Believe*. I offered him a copy to read if he wished. He held it for a moment, scrutinized it carefully, and then handed it back to me. I've blown it, I thought. I have done the wrong thing by raising this issue. But then, with a smile, he asked, "Would you kindly autograph it for me?"

CHAPTER 36

LETTER

The immediate hurdle I faced was asking Pope Francis, the official custodian of the Shroud, to consent to access to the Shroud of Turin so that new samples could be taken for DNA testing. Esteemed and deserving scientists working on the Shroud had tried and failed. What made me bold enough to think that I might succeed? The answer was simple. It was the Buenos Aires Case itself.

When Bergoglio was Archbishop of Buenos Aires he wanted the case investigated. And it was. I have done the work of that investigation, of obeying that request, for close on 20 years. In order for that request to be completed the next stage requires permission from the Pope to test individual cells on the Shroud for comparison. It seemed to be a reasonable and logical progression which I felt certain would be compelling enough to persuade Pope Francis to grant the permission to proceed.

I spoke to Padre Alejandro Pezet, the priest from Buenos Aires who was central to the events of the Eucharistic miracle in 1996. He is a personal friend of Bergoglio from his time as the Archbishop of Buenos Aires. Padre Alejandro was enthusiastic about all the new developments in the investigation and encouraged me to write the letter. He went even further. He made an offer: If I emailed my letter to the Pope to him he would personally make sure it got to the Pope. He said he had contacts that could make that happen.

This is the letter:

His Holiness, Pope Francis

Vatican City, Rome, Italy

RE: INVESTIGATION EUCHARISTIC
PHENOMENON, BUENOS AIRES 1996

Dear Holy Father,

When I met you on 17 March 2006 it was to deliver the results of the scientific testing on the communion host from the Santa Maria Buenos Aires parish that spontaneously transformed in 1996. With me was Padre Alejandro Pezet, the parish priest involved in the discovery of the transformed host.

Included in the report were the findings of an esteemed forensic pathologist and heart specialist I engaged in New York. With no prior knowledge of the source of the sample I supplied, he concluded emphatically that what he was seeing was human heart tissue from the left ventricular wall replete with white blood cells indicative that the person had suffered traumatic injuries. Defying any scientific explanation the sample which had been in water for three years showed no degradation.

My investigative work on this case has continued for twenty years. All of it has been documented and filmed. As science has advanced so have the opportunities for further testing. These include genetic testing.

To the surprise of multiple scientists in different countries, standard DNA tests have all mysteriously been unable to extract a genetic code from the nucleus present in the cells of the sample. Fortunately there has been a very recent scientific breakthrough in technology for DNA testing.

Menarini Silicon Biosystems, a scientific laboratory in Bologna, Italy has invented a system to successfully isolate a single white blood cell from a sample and to genetically analyse it with microscopic precision. In 2016 they were able to isolate and examine white blood cells from our sample and do successful mitochondrial DNA testing within the non-coding region of the human genome with Next Generation Sequencing technology. Informative data regarding the maternal heritage of the person was revealed.

It seems to me that the next logical step in the investigation is to compare the mitochondrial DNA data that was obtained from testing the Buenos Aires case with the same mitochondrial DNA testing that could be done on single blood cells from the Shroud of Turin.

Dr Francis Collins, an esteemed member of the Pontifical Academy of Science, renowned for leading the International Human Genome Project and his expertise in human genetics would be perfectly qualified to undertake the investigation. But of course any access to the Shroud of Turin comes only with your permission.

I humbly ask you for this permission, and if granted, that you invite Dr Francis Collins to become involved.

I look forward to hearing from you.

Yours sincerely

Ron Tesoriero

6 December 2019

On 27 December 2019, Padre Alejandro called me and said, "I have spoken to the Secretary of Pope Francis. The Secretary said that he had received my letter and had placed it on the Pope's desk."

I have not yet had a reply from Pope Francis. Jesus told Katya Rivas emphatically that comparative tests on the Shroud will be done. No testing can happen without a Pope granting permission. That may be the current Pope. Or perhaps a future pope.

CHAPTER 37

COINCIDENCE

*Of all the words in the Old Testament, '
heart' ranks third in the list of those most often written.
In first place is the word
'Lord' and in second place, 'God'.*

I often hear people of faith remark that there are no coincidences with God. What seem like coincidences are often part of a divine plan of God. 'Coincidence' Albert Einstein once said, 'is God's way of remaining anonymous.'

How coincidental is it that Pope Francis is pivotal to the Buenos Aires communion host story? In 1999 he, then Archbishop Bergoglio, wanted it investigated. Some 20 years later to further that investigation access to the Shroud is needed. The only person who can give that consent is the official custodian, the Pope, and again, coincidently, the same man now happens to wield that authority. But I see a further coincidence, or as one might say, a plan of God, in that the same man is not only at the centre of this amazing Eucharistic miracle of Buenos Aires story, one of the most significant events in the history of Christianity, but is also the administrative head of the Catholic Church. After analysis of the facts and scientific findings it is easy to see all these coincidences as God stepping into our physical world in a very powerful way to tell us something important about himself:

- that he has power over the natural world;

- that he has the power to bring into being human life;

- that he has the power to create living white blood cells and living human heart tissue from non living matter spontaneously;

- that he is the Creator and our Creator;

- that from the very beginning man was created and did not evolve from lesser forms of life;

- that Jesus makes himself alive and present in the consecrated Communion Host;

- that the Shroud of Turin is his true burial cloth, stained with his blood that was shed from severe beatings and from his crucifixion;

- that he has left the signs of his suffering not only on his burial cloths but also in the communion host of Buenos Aires and his bleeding statue in Cochabamba;

- that Jesus has permitted us to verify through science and through genetic studies that he is really alive and present in the Holy Eucharist;

- that he has provided the potential to discern, through ancestral DNA studies, that he is the promised Messiah from the line of King David;

All of this is a powerful ecumenical package which has fallen into the hands of the leader of the Catholic Church and if it was widely publicised could change the direction of our predominantly atheistic world. More specifically, what has happened has the potential to revive faith in the Real Presence of Jesus in the Eucharist.

The context in which the samples from the Buenos Aires Case arose is relevant. The transformation of the bread of the communion host into flesh and blood happened in a Catholic Mass. The Catholic Church proclaims that at the moment of consecration the bread and wine becomes 'truly, really and substantially the Body and Blood, Soul and Divinity of our Lord Jesus Christ.' The bread and wine still look the same but through a mysterious process called transubstantiation the substance of the bread and wine is changed into a truly present Jesus, as present as if he were actually standing there in person. This is called the Holy Eucharist.

When the priest offers up the bread and wine at the moment of consecration he does so following the command Jesus gave at the Last Supper.

The Gospel of Luke provides this account (22:17-20):

"On the night before he was betrayed, while at supper with his disciples ,Jesus took bread, said the blessing, broke the bread and gave it to His disciples saying, "Take this, all of you and eat it. This is my Body which will be given up for you." In the same way He took the cup filled with wine. He gave thanks and giving the cup to His disciples said, "Take this all of you and drink from it. This is the cup of my blood, the blood of the new and everlasting covenant. It will be shed for you and for all so that sins may be forgiven. Do this in memory of me."

This belief in the Sacrament of the holy Eucharist is the central and most important teaching in the Church and has been since its very beginning. It is a belief of immense and magnificent ramifications.

What makes the Buenos Aires case a more powerful statement and aid to understand the Eucharist than any of the other more recent cases of Eucharistic miracles comes from what Dr Zugibe said as he looked at the sample under his microscope.

Dr Zugibe spoke about it being from the heart of a living person who had suffered traumatic injuries like he had seen in cases where a person had been severely beaten around the chest.

"I am looking at the snapshot here of a living person. The injuries were inflicted on this person some 3 days beforehand."

I questioned theologian Monsignor Micek about this.

Me: "If Jesus suffered traumatic injuries from having been tortured and beaten on the Friday, the day of his crucifixion and death, how, in theological terms, does one explain that Dr Zugibe is seeing the heart of a living person that had suffered traumatic injuries three days before?"

Mons. Micek: "Catholic teaching maintains that the communion host we receive at Mass is a memorial of the Passion, Death and Resurrection of Christ and that at the moment we receive communion it reflects the moment of the Resurrection."

That Jesus is really present and alive in the Eucharist is perhaps the most difficult concept of all Catholic dogma for the rationalist of today to accept. Many Catholics, like Protestants, take the view that Jesus is only symbolically present in the Eucharist. Perhaps it is for this reason that God has given this Eucharistic Miracle, along with the others in the last decade, to encourage a return to faith in his Real Presence in the Eucharist.

I have given many talks on the Buenos Aires case, in different parts of the world and have been surprised by how positively people have reacted to knowing about the facts and findings. Many have written telling me that they have either come back to the faith or that they have gained a renewed understanding of Jesus being present in the Eucharist. In those talks I have emphasized that there was 'heart' in the Eucharist. I drew upon the impressive teachings Jesus gave to Katya in messages about the Eucharist and his *'Eucharistic Heart'*. Jesus has said to Katya, *"I have placed my heart in the Eucharist"*. Jesus has confided the same to some other saints in history. Notable are St Faustina Kowolska and St Margaret Mary Alacoque, who is known for her establishment of devotion to the Sacred Heart of Jesus. Drawing from these teachings and the implications of the scientific findings in the Buenos Aires case I always end my talks with this note of encouragement:

Me: "Jesus said after his Resurrection, and before his ascent into Heaven that he would not leave us as orphans. He would remain with us until the end of time. He chose to remain with us in the Eucharist. He offers this opportunity not only to Catholics, but to every person on the planet. The next time you stand in line waiting to receive Communion think about what you are going to receive. Jesus. And who is this Jesus? He is the creator of the world. He is your creator. He is the one who will judge you upon your death to determine whether you are eligible to enter into Heaven. He will then welcome you as royalty to take your place in his kingdom, an indescribable paradise, that will last forever. Yet that Jesus, that king, humbles himself to come to you hidden in the communion host anxious to place his loving heart next to yours as a foretaste of being with him in Heaven. This knowledge has changed my whole approach to Communion. I hope it will be the same for you."

Most Catholics have a respectful but poorly informed understanding of the Eucharist. Can you imagine the impact if the clergy and Catholic educators were encouraged to offer this same reflection on the Eucharist to those in their spiritual care?

One high profile person of faith made a dramatic decision upon learning the facts and findings of the Buenos Aires case. He was, before his conversion, personal chaplain to her Majesty, Queen Elizabeth 11, Queen of England and the head of the Anglican Church. In December 2019, newspapers were reporting that a prominent Anglican Bishop, Dr Gavin Ashenden, and Chaplain to Queen Elizabeth, had converted to Catholicism. One of the reasons he listed for his decision was the impact of the facts and scientific findings of the Buenos Aires case. One publication of the story quoted Dr Asheden saying, "One of the discoveries that affected me deeply were the Eucharistic miracles. There are very many, but the most recent are the most interesting because of the science." He then proceeded to articulate the facts and findings of the Buenos Aires case. In conclusion he said, "In my judgment, it was looking like the Catholic Church had been right all along about Mary, and about the Mass."

In effect, someone deemed worthy to be the religious advisor to Queen Elisabeth, the head of the Anglican Church, is declaring that Protestants are wrong to reject Catholic belief about transubstantiation and the Real Presence."

https://www.premierchristianity.com/Blog/Former-Queen-s-chaplain-Gavin-Ashenden-Why-I-m-converting-to-Catholicism

A well known Anglican Bishop has had a dramatic conversion to Catholicism because he happened to find out what I had done and documented regarding the scientific tests and findings. I would not have been in a position to have done this work unless Pope Francis, when he was still an Archbishop, had decided he wanted an investigation in the first place. Just imagine how many more conversions will occur when Pope Francis is ready to speak about the results? What's more he has the assurance of Jesus himself that when he permits the blood comparison with the Holy Shroud the scientific results will provide what is needed for the rebuilding of the Church.

On 6 September 1996 Katya was at home in Bolivia when she wrote down this message dictated to her by Jesus:

"Indeed, my Blood will be compared. Did I not tell you so before? It is necessary to compare it with the Holy Shroud. It is necessary to give noble weapons to the people of my Church in order to preserve the faith of its members in these times when the devil of materialism poisons its blood. Before the glorious time comes when my Church, by then united, declares me as its leader and leaves aside selfishness, the Scripture will have to be fulfilled"

Little did she know that thousands of miles away, in a different city, Buenos Aires, in a different country, Argentina, on that very same day an extraordinary event was unfolding: a communion host had turned to heart. Was this just coincidence? I can only deduce that if Jesus was permitting his blood to ooze from a communion host in Buenos Aires and on the same day he says that his blood will be compared with the Holy Shroud, the two matters must be related. Integral to both is Pope Francis. The Buenos Aires case is his, and the decision to compare the blood with the Holy Shroud is his.

The stakes couldn't be higher: the preservation of the faith of the whole Church in a time of crisis is virtually assured if Pope Francis embraces the opportunity that now lies before him. The letter is on his desk. The authority is in his hands.

CHAPTER 38

I HAVE DONE MY WORK

On 5 October 1999, three years after the mysterious transformation of the consecrated communion host in Buenos Aires, I took the first sample for testing. Even though the witnesses I filmed explained the circumstances under which the transformation had occurred, I had no idea what the testing would reveal. The dark gelatinous substance in the test tube could have been anything.

On my way to the laboratory in San Francisco, on 20 October, I had a meeting with Katya Rivas. I explained what tests we were planning to do and showed her the Buenos Aires samples concealed in the test tubes. She reached out, took them, drew them close to her chest and closed her eyes. She put the samples down on the table, took out her notebook and began to write. Jesus was speaking to her:

"My dear ones, pray while everything is carried out, on this first day so important in the history of humanity. When the world is on the path to perdition, I come one more time. I, the King of Glory with my suffering, cry out to man of my love for him, and my real presence among you.

Thank you RC and Ron for taking on this case, this work. Do not be concerned if you come across contradictions or setbacks, if I am going to be guiding you and your interests are also mine."

Now, after twenty years of working on this case, after achieving astonishing scientific results, I am beginning to appreciate the magnitude of that message. At the time I had no idea of its significance. If it is truly from Jesus, which I believe it is, then it is worthy of the type of reflection one can only truly engage in after time has passed.

Jesus starts the message asking that we *"pray on this first day so important in the history of humanity."* This was the first day of the beginning of the scientific investigation of the Buenos Aires Case. In essence it is Jesus, God himself, pronouncing to the world that the outcome of the Buenos Aires case, of this work we will do on the testing *"will be so important in the history of humanity."* He calls it *"this case"* but then adds *"this work"*. He knew something I didn't know at the time: that I would continue my work long after these initial scientific tests were done. This would be merely the first of many laboratories I would deal with, of many scientists I would interview, of many scientific textbooks I would read, of many countries I would have to travel to for another twenty five years.

If it is important, as he says, then presumably it has to do with putting something right in the recorded "history of humanity." When we write of the history of anything we start from the beginning, from its origin. Thinking about the history of humanity leads inescapably to one of two conclusions about its origin: either man was created, or man wasn't created. If man was created then a Creator did the creating. It was an act of special creation. Nothing 'evolved' to create human beings. If he wasn't created, as the majority in all educational, government, media and scientific institutions believe, then his existence is a result of billions of years of evolution. In this view man came into existence after one species evolved into another, each more complex than the next, until eventually primates evolved. Mankind is, essentially, the pinnacle, a species of highly evolved ape. From the revered Encyclopedia Britannica one reads, 'The evidence is overwhelming that all life on Earth has evolved from common ancestors in an unbroken chain since its origin.' (https://www.britannica.com/science/life/Evolution-and-the-history-of-life-on-Earth)

Proceeding from this understanding of humanity science attempts to fathom all aspects of human biology and behaviour.

The belief that humans are a product of an evolutionary process has made it easy to remove God from the very first chapter in the history of humanity. In coming to set right that history of humanity through the results of the Buenos Aires case, Jesus announces himself as the

"King of Glory." These are not everyday words. But they are seen in the Bible. What do they mean?

'Glory' has many shades of meaning. It sometimes describes a physical visible phenomenon, some manifestation of divine grandeur, truth, goodness or power. It denotes a judgement of great worth and highest honour. Often it invokes the majesty of God as the Creator of a magnificent universe, as the cause of everything that exists from nothing. Everything which he has willed into existence is in some measure a reflection of himself. By reason alone God can be known in his material creation and in the immaterial laws which govern it.

Whereas the word 'glory' is found throughout the scriptures the title, *'King of Glory'* is found only once. It was conferred on God by King David in Psalm 23. He declares that the earth and everything it contains and all who live in it are subject to him because he created it. We are his because he created us. As the Creator he claims absolute authority. He is the *King of Glory* with the right to rule over everything. The debate over the origin of life is not merely academic or scientific. The contest of ideas is not simply about how life came into existence or how the universe was made. It is about who is in charge. If God is not our Creator then he cannot be our King.

The power to create life in the Buenos Aires Eucharistic host is the selfsame power of the King of Glory described in the psalm. It is the power of Creation. He wants us to know that he has intervened and is permitting Science to confirm that what happened when a piece of bread became flesh can only be attributed to him. It was an act of creation. Only he can create. The message Jesus gave declares that he is in charge. He is the creator and therefore he is the king. As the *King of Glory* he is asserting his existence and correcting the errors of the origin of life story being told nowadays, errors which have so grievously distorted human history and corrupted human society.

For those who think God created life as a mysterious biological ingredient in the beginning but then retreated and let evolution run its course there is no longer any intellectual comfort. Multiple cases of mysterious Eucharistic transformations in which scientific evidence of human life suddenly emerges out of nonliving bread proclaims a very engaged and active God. He is not merely watching from afar. He is involved. He intervenes; in our very own times.

In the 1996 intervention in Buenos Aires he is adding something important to the *"history of humanity"*. The communion host transformed into living human heart tissue. It was found to be infiltrated with living human white blood cells. Those white blood cells, as cells, are the basic units of life and they are clear scientific evidence of life. They have come into existence spontaneously from non living matter.

Science today has no idea how life started or how the first cell of life came into existence. Many scientists have put forward theories as to how it might have happened, but those theories are exactly that: theories. Famous Oxford evolutionary biologist, Professor, Richard Dawkins was pointedly asked by Ben Stein in a televised interview for a documentary called *'Expelled'*, " How did life begin?"

He responded unhesitatingly, "No one knows how life began ."

In the Buenos Aires case we have, in our physical world today, those physical living white blood cells which were tested. We know the very particular circumstances in which they came into existence: a Catholic Mass which is a liturgical rite of worship of the God of Abraham, Isaac and Jacob, the God of David. In the Buenos Aires case God is revealing, through science, that he is the Creator of life, and that means he has the power to bring life into being that was not there before.

Forensic pathologist , Dr Zugibe was clear in his report that the material was human cardiac muscle from the left ventricle wall close to a valvular area. Heart tissue spontaneously came into existence from non living matter. The physical evidence of this exists today, and the circumstances of how it happened are known. The chain of evidence has been meticulously documented. There was no evolutionary process that preceded the coming into being of the human heart from which that tissue came.

An evolutionist typically selects a complex organ of the human body and compares it with what that organ looks like in other simpler creatures on a branch lower down on the evolutionary tree of life. The assertion is that a human heart didn't just turn up the way it exists today. Instead it developed in stages over millions of years. The basic version goes something like this: First there was an invertebrate stage with no heart as such, just a muscle system pumping blood. Then it evolved to a rudimentary form of heart like that of a fish. Then it progressed to a three-chambered frog heart, then a three-an-a-half chambered turtle heart then to a bird heart and finally to a four chambered human heart.

How do evolutionists account for the evidence in the Buenos Aires spontaneous host-to-human heart case. There were no billions of years of evolution. No prior invertebrate, no prior fish, no frog and no turtle. There were no intermediary steps. There was no evolution whatsoever. The essence of Evolution lies in the conviction that all life forms on earth today are a product of an evolutionary process. All means all. The theory does not allow for even a single exception. If there is one exception to the rule, the theory must fall. Charles Darwin himself said that in the event that such an exemption existed his theory would absolutely break down.

The facts and findings of the Buenos Aires case clearly present that exception. Looked at purely in factual and scientific terms I am able to deliver this death blow to evolution in the form of an argument that has never been presented before. Now, because of the Buenos Aires case, it is. No-one can successfully raise the evolution argument against Jesus ever again. He has, with the elegant simplicity of the Buenos Aires case, effectively shredded 150 years of Darwinian evolutionary argument against his existence. The biblical account of the creation of man in Genesis is made relevant again. No longer can it be derided as a fairy tale.

In that message Jesus gave on 20 October 1999 he said he comes again *"with my suffering cry out to man of my love for him, and my real presence among you."* It occurs to me now that he was telling us in advance what we would find out about him from the scientific testing. Forensic pathologist, Dr Zugibe, identified the material taken from the Buenos Aires host as heart tissue. The heart is regarded as the symbol of love. The presence of white blood cells in the tissue, indicated to Dr Zugibe that the heart tissue was from a living person who had suffered traumatic injuries. That type of suffering was to show up not only in forensic pathology testing but also later in the mitochondrial DNA testing.

His words, *"my real presence among you"*, reinforce the Church's teaching that Jesus is a living real presence in the Eucharist until the end of time.

Not to be forgotten is the potential for a whole new chapter that may be written into the *"history of humanity"* which has to do with the Virgin Birth, the ancestry of Jesus as a descendent of King David and even possibly further back to the very beginning with Adam and Eve.

I wrote to Pope Francis requesting permission to scientifically compare the blood on the Holy Shroud with all of my test results. I suggested the whole exercise be undertaken by the world famous geneticist and member of the Pontifical Academy of Science, Dr Francis Collins. No-one more qualified could interpret the complex genetic data on the ancestry of Jesus from the Italian lab results that might be replicated in similar testing on blood on the Holy Shroud.

If they are compared, which Jesus has said will happen, and if there is a match, then the consequences for humanity will indeed be enormous. The white blood cells from Buenos Aires communion host were from a living person. The blood on the Holy Shroud is from a long dead person, 2000 year long dead. If the two match, then the blood from the person who died from crucifixion is from the same person who is present and alive today in the communion host. A dead person and a living person have the same DNA fingerprint?

He who was once dead is now alive. The New Testament records that Jesus proclaimed that he would rise from the dead to confirm who he really was. Science is now poised to make such a declaration. Such an outcome could conservatively be categorized as *"important in the history of humanity."*

On the 6th November 2006, Time Magazine ran a leading article on a debate between two eminent professors, Francis Collins and renowned atheist Richard Dawkins. Both had recently published books. Collins's book was called 'DNA-The Language of God' and Dawkins's was called 'The God Delusion'. Time characterized it as a boxing match – God vs Evolution.

The spirited exchange reveals the differences in mindset on the issues that will confront Dr Collins should he ever be asked by Pope Francis to compare the Shroud with the Buenos Aires host case.

TIME: Dr. Collins, the Resurrection is an essential argument of Christian faith, but doesn't it, along with the virgin birth and lesser miracles, fatally undermine the scientific method, which depends on the constancy of natural laws?

COLLINS: If you're willing to answer yes to a God outside of nature, then there's nothing inconsistent with God on rare occasions choosing to invade the natural world in a way that appears miraculous. If God made the natural laws, why could he not violate them when it was a particularly significant moment for him to do so? And if you accept the idea that Christ was also divine, which I do, then his Resurrection is not in itself a great logical leap.

TIME: Doesn't the very notion of miracles throw off science?

COLLINS: Not at all. If you are in the camp I am in, one place where science and faith could touch each other is in the investigation of supposedly miraculous events.

DAWKINS: If ever there was a slamming of the door in the face of constructive investigation, it is the word miracle. To a medieval peasant, a radio would have seemed like a miracle. All kinds of things may happen which we by the lights of today's science would classify as a miracle just as medieval science might a Boeing 747. Francis keeps saying things like "From the perspective of a believer." Once you buy into the position of faith, then suddenly you find yourself losing all of your natural scepticism and your scientific—really scientific—credibility. I'm sorry to be so blunt.

COLLINS: Richard, I actually agree with the first part of what you said. But I would challenge the statement that my scientific instincts are any less rigorous than yours. The difference is that my presumption of the possibility of God and therefore the supernatural is not zero, and yours is.

I am sure that once Francis Collins is made aware of the facts and findings of the Buenos Aires case it will provide him with the means to destroy Dawkins' argument.

Jesus ends his message giving me these words of encouragement: *"Do not be concerned if you come across contradictions or setbacks, if I am going to be guiding you and your interests are also mine."*

Throughout the two decades of working on this case I have had setbacks, one after another. Each time I felt I could go no further with the case. But in every instance, at the point of exasperation, a coincidence would throw a light on my path and allow me to take another step forward. Now, so near the finish line of this journey, and again I am confronted with another set back. I have had no reply from Pope Francis to my request for his cooperation in testing the Holy Shroud and comparing the results thereof with the Buenos Aires case. So much depends on his cooperation. But I am not concerned. I am confident that the official guardian of the shroud will cooperate. The one who has been guiding me throughout, the King of Glory with his Almighty power, can and will make it happen because he has said to me *"your interests are also mine"*. And his interests are way beyond mine; his are of eternal magnitude. He says that the *"world is on the path to perdition"*. Perdition means more than the secular notion of failure and disaster. It carries with it a grave finality, a point from which there is no return and only tragic eternal loss.

There is so much to gain and so much to lose.

CHAPTER 39

MIKE

Not everybody is honoured by heads of state, a requiem Mass offered by the Archbishop in a Cathedral, and headline obituaries in major newspapers when they die. My longtime collaborator and friend Mike Willesee was. On 1 March 2019 all across the Australian media landscape people were paying tribute to one of their finest. Re-runs of groundbreaking interviews with celebrities and Prime Ministers were accompanied by accolades for 'the father of Australian investigative journalism.' Fellow journalists wrote glowingly of the influence he had on them. Kerry O'Brien, the former long-time presenter on Australia's national broadcaster, "He was an absolute trailblazer on Australian television. I regarded him right through my career as the benchmark. He was quality and class from the outset. A very sharp mind and a great instinct for the right question." Another prominent Australian journalist, Ray Martin, said, "He set the bar in Canberra for television journos and half a century later no one has jumped over it, although there have been some worthy attempts."

St Mary's Cathedral was packed for his funeral with well known faces, the movers and shakers, some from media, film, television and publishing, others from the political and sporting world. Sydney's Archbishop Anthony Fisher read a tribute to Mike from the Prime Minister of Australia, the Honourable Scott Morrison. Many in the pews were, for the first time I imagine, gazing in awe at the splendour of the stained glass light and impressed by the solemn beauty of the ancient Catholic liturgy. For them God, Church, and religion was alien territory. It soon became obvious that this great icon of the Australian media was no stranger to God. And he was not afraid to let it be known.

In an interview he gave to Sydney's Catholic Weekly newspaper, he said, "I think God and religion are just so politically incorrect at the moment. Nobody wants to know. You can be laughed at just for saying you believe in God. The work I'm doing now has been difficult. I'm constantly reminded that most journalists would say this is ridiculous. But I've stuck with this story and I'm making progress. It shows the truth of God in the Eucharist. The truth that God is alive in our world and that his hand moves."

I was very aware that there was a whole other life of Mike which was completely unknown to his peers. What if they had recorded some of our most profound and mysterious moments? What if they saw what we saw and heard what we heard? Would their indifference to God and religion still be as strong?

We spent so many hours sitting next to each other on planes in order do this work that I unfortunately remember only a few of our many conversations. One stuck in my mind:

Mike: "Did you ever think that I would join with you in doing this work?"

Me: "I had no doubt that you would say "yes" to me."

Mike: "Why do you say that?"

Me: "I could see from what I had observed of you in your role as an investigative television journalist over the years that you were out to find the truth. You would not stop until you found it. And when you did you were not afraid to tell it publicly even if it meant some personal discomfort or the loss of friendship. Even of the prime minister of the country."

In Mike's spiritual biography which was published after his death he wrote. 'I chose journalism and my life came alive. I thought I'd be a good reporter because I was a good writer. As it turned out, I wasn't that good a writer, but I did already possess some of the qualities that would hold me in good stead for 50 years as a journalist. I was genuinely curious – I always wanted to know the truth. I was dogged – I hated going back to the office without the answers. And I was ambitious – I always wanted to achieve something. For me, it wasn't enough to get the story – I wanted to make a difference to the situation. And I was brutally honest in the way I reported the facts and told stories – I hated bullshit and deceit… My reputation was as a journalist who sought truth, uncovered deceit and asked hard questions to find hard answers.'

In the work we were doing there were many hard questions to be faced. How could a plaster statue of Christ shed tears and human blood? How could a communion host transform into living human heart tissue? But perhaps the hardest question, and the one which most tested his reputation concerned the mystic Katya Rivas. Was she for real? Were her experiences genuine? Was God dictating to her? Can we really believe all this?

I first introduced Mike to Katya in Cochabamba, Bolivia in August 1998. Mike came with me specifically in order to form his own assessment of her. He spent a week or so observing her going about her daily duties. On one of them she said that the Virgin Mary was going to appear to her in the nearby town of Candilaria, in the very same place where she had her first supernatural experience with the Virgin Mary some years before. So we accompanied her and I filmed the events on the day.

I asked him about it in an interview:

Mike: "I was sceptical. I was sitting on the fence and I didn't believe, and I went there not believing. After 35 years in journalism, I think I know something about people, and I wanted to see for myself what sort of people these were. Is it ego? Are they trying to start a cult? Is it money? Like we see for so many. Are they trying to make money out of it? So I watched for some days and didn't see any example of this. Instead, I saw humility and prayerfulness so I was quite inclined to believe that I was seeing a matter of truth. But when Katya dropped to her knees and I happened to be watching as she dropped, and, stunned, fell back, and it was certainly very impressive, and very few actresses could have done it, but it still didn't prove to me. I couldn't see anything. All I saw was a very credible performance by Katya, but I still had an open mind. But during the day, which was a fairly warm day, and a long day, we travelled a long distance that day, I noticed she kept wiping her hands, and I wondered what this was about. Most people didn't seem to notice. But I noticed that Father Renzo was sneaking her tissues while we were praying, and she was giving back the wet tissues. It seemed to be bothering her all day, this wetness from her palms, which seemed beyond what you would expect of perspiration. She wasn't perspiring from the face or, if she was perspiring from her body it wasn't showing in her clothes. Late in the day, we went to a church, we were going on the way back to Cochabamba from the country, and I saw her getting almost irritated, because she was running out of things to wipe her hands, so I passed her a cloth, and almost absentmindedly she wiped her hands and just passed it back to me, a little thank you, and I smelt it, and I was completely astonished. It was a fragrance that I don't believe was man made, and I don't believe a fragrance can come from any part of the body, after a hot day's traveling. It was beautiful, it was a fragrant oil, and I just cannot think of any rational explanation for that, and that was the first time for me, I thought something supernatural is happening here."

Mike had witnessed the inexplicable exudation of a perfumed oil coming from Katya's hands. It had no natural explanation. And for him it pointed to a spiritual reality.

From those same hands Mike witnessed something quite startling that also pointed him towards a spiritual reality. Katya prophesied on film, three months beforehand, the time and place, that her hands and feet would bleed and that she would suffer the stigmata of the Passion of Jesus. And it all came to pass exactly as she said. We accompanied Katya at different times to different parts of the world where we witnessed mystical phenomena associated with her. Numerous times we watched her write messages that she said were being dictated to her by Jesus. Occasionally she would write profound words that were relevant to matters we were tackling.

I remembered being with her in the town of Lanciano, Italy, the site of the first significant Eucharistic miracle which occurred in the eighth century. As we watched she wrote pages of important teaching on the Eucharist which impressed the theologian travelling with us.

"Dear Children

Here is one of the biggest moments of love that I have given to incredulous man – that My body and My blood materialised to give you not only the mystery of faith, but faith coming from the mystery, to become reality before the eyes of humanity. Who knows about this place of love? Those who want to know it; the ones who believe in My Presence – My Real Presence in the Eucharist, the simple souls, the simple ones, those to whom I am calling. Blessed are those who are poor in spirit, the simple, the poor ones in front of evidence of God who loves infinitely, not only to the point of giving His life for man, not only to the limit of giving them His Blood, and His Body, but even beyond every limit where human crosses the frontier of the Divine to make it possible that the Divine becomes human before the eyes of His beloved creatures. Do you really know in what place you are? The place where you are is like the door of heaven where the Holy Trinity becomes flesh and that flesh remains throughout the years."

In October 2000 we all visited Turin for the public exhibition of the Holy Shroud. We walked past a curated display of occasions throughout history when the Shroud had been publicly displayed. Eventually we arrived at the relic itself. There. In our presence, whilst kneeling before it Jesus dictated some explanations to Katya:

"If you had taken note of the reproductions that come before this exposition you would see that it was handled by many people who throughout history exposed it publicly. Imagine how many other times it was exposed privately. Any other cloth like this, which even only being folded would have disintegrated already or would have become stained and by today it would have become a dirty and ordinary piece of cloth. Another aspect that I want you to observe is that the part of the body that is exhibited on the front of the shroud, and on the back, has no important burn marks that would have needed to be mended; from every point of view impossible."

Jesus was making reference to the amazing fact that the relevant sections of the body image were not damaged or destroyed by a fire which destroyed other parts of the shroud hundreds of years before.

The subject of the Shroud has come up a number of times when Katya received dictation from Heaven in the presence of Mike and myself.

Many times, in the context of Jesus issuing guidance and instructions on our investigations and documentaries the subject of the Shroud has come up. He encouraged us to make 'Blood of Christ', a documentary on Eucharistic miracles including a comparative analysis on the Shroud of Turin. On one particular occasion in Mexico in 2001 Mike asked a question which sought further clarification on the instructions that Jesus was giving. Katya began to write out the answer Jesus was dictating. I filmed Mike watching Katya writing out the answer to his question. The person next to Mike says, " Mike ,can you imagine that God is talking now like with Moses. He is talking now as he did with Moses 3,000 years ago. This is like Sinai here."

Another time, in June 2001, while sitting with Katya at lunch in a restaurant in Sotto Grande, Spain, discussing the progress of our investigation of the Buenos Aires case, Jesus joined the conversation through Katya:

"The first time you had in your hands the host in Buenos Aires you did not imagine the path you had to follow and the point that you are at now. Now you have to show to others the truth of the testing. And that I say especially to Ron. When you began the project in Argentina you would never have suspected what I was planning to do , what I really wanted of the work from you, because if I had told you about the results you now have , you would have thought that Katya was crazy. How can I tell you step by step what you have to do, even when I give you all the means. You don't take enough care. What happened to the samples Ron?" (I cowered in shame at this revelation because I had not told anyone, and certainly not Katya, that I had misplaced one of the samples from the Buenos Aires case. No one knew. But Jesus did and was raising it as an example of my not taking enough care in his work.) *"I know how much you have invested in this project. I know about the difficulties and the sacrifices that you have to live through. You have to plan everything very carefully. I cannot tell you everything because our common enemy is trying to discover my plans and to interfere with them.*

The only thing I can tell you is that I am the owner of time and of everything you possess. Tomorrow you will be an important part of those who have helped my Church to give back the faith to man."

On 7 February 2001, in Merida Mexico I filmed Katya Rivas writing a message dictated by Jesus about the investigative work of Mike and myself. Someone present exclaimed , "God is talking now as he did with Moses 3,000 years ago."

On 10 June 2001, in Sottogrande Spain, I filmed Katya writing a message being dictated by Jesus adressing our work in the investigation of the Buenos Aires case.

On 20 August 2017, Merida Mexico, Mike began a series of interviews with Katya about her experiences.

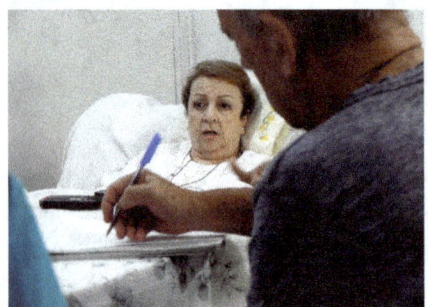 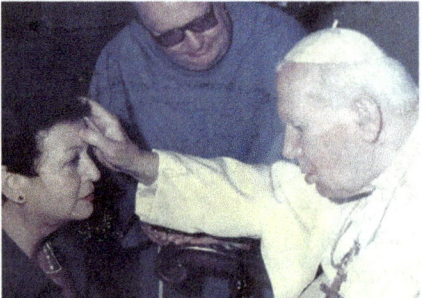

On 22 August 2017, I filmed Katya telling Mike of her bilocation experience with Pope John Paul 11.

On 23 August 2017, Merida Mexico, Mike said his last goodbye to Katya.

CHAPTER 40

THE FINAL INTERVIEW

After Mike had been diagnosed with cancer and was given only months to live, he set upon completing a book that he was writing on his conversion and on his involvement with Katya. He wanted one last series of interviews with her, probably taking place over a whole week, to complete the book. He needed a full week with her. But his diagnosis also meant multiple medical appointments which he couldn't miss. He had to be in Australia on the first day of September. He asked me to contact Katya to line up a week before September. There was a small sliver of time in which he could manage to travel to Bolivia. It would have to be the last week of August. I made the bookings.

On the flight we discussed the subject matter for the interview. As a final reflection I said, "Do you realise that we have been privileged to have had, almost exclusively, the opportunity to have documented over 20 years of Katya's remarkable story; someone who will perhaps go down in history as having had mystical experiences equivalent to some of the Church's most famous saints; experiences that could influence the course of history? That privilege I think comes at a cost. You as a journalist and me as a lawyer have a duty to tell that story. None of us, not you nor Katya nor I, is getting any younger. We are in our twilight years. It seems to have been a pattern whenever we have ventured out on a journey in this work we come home saying to ourselves, from what has happened on this trip, we were meant to have gone there. I would not be surprised if that happens again to us."

We arrived in Merida in Mexico and sat down for the first interview with Katya. She said, "At the beginning of August I was planning to go to the United States for a meeting. Jesus then told me, *"Do not go yet. Stay in Merida, I want you to keep the last week of August free."* He did not tell me why. But now I know. It was so that I could be here for you to have me for the week you wanted."

There were three areas that Mike chose to highlight in his book:

1. Her experiences of bi-location. There are recorded corroborated testimonies of some saints, like Saint Pio, having the facility to be present in two places at the same time.

2. Her interaction with Pope John Paul II.

3. Acceptance of her messages by the Church.

Katya spoke and we listened and recorded.

"In 1998, I went to a conference in Rome to which a number of my group had been invited. My spiritual adviser, Father Renzo Sessolo, was invited also but was delayed on his arrival because he was travelling with our Archbishop, René Fernandez. When we finished dinner one night, we gathered so we could get ready to pray, but Our Lord said to me, *"Get ready because you will meet the Pope tonight."* I did not know when or how that would happen. And the only people I told were my two closest friends there. Then I fixed my hair and held one of my recent notebooks that I had been using to take dictation from the Lord. I had it next to my chest. Then the Lord said to me, *"Lie down."* I lay down on the bed and began to pray and I told my friends, who were sharing the room with me, not to do anything, just to pray, even if I go flying out the window like Superman."

Mike interrupted her at this point because it sounded like she was taking a bit of poetic licence.

Mike: "Did you actually say, 'Even if I fly out the window?'"

Katya: "Yes."

Mike: "Was that a joke?"

Katya: "No. I did not know how I was going to see the Pope. I just knew that I had to lie down on the bed. I did not know if it would be a dream or if the Holy Father was in fact going to fly in. I did not know what bilocation was.

I went to sleep and when I woke up, I was in a different place. I saw the Holy Father, John Paul II, from behind. Without turning around, he said, "Is that you, Katya? I was waiting for you, Katya." He had his rosary in hand and was saying the rosary and then he turned around and I fell to my knees and I started crying. He held out his hand and I kissed his hand. I asked him to forgive my sins because I had judged him badly because I thought a Pope had to be like Pope Pius XII. I told him everything and then he said, "Don't worry." And then he saw that I was bare-footed, and he kicked some slippers over to me. "You forgot your slippers."

When he gave me his slippers he said, "Many have seen these slippers but no one else has ever worn them before."

He then said, " I was waiting for you. You should have come with a priest." He then appeared to be thinking and he touched his head as if in thought. He said, "Oh, he is flying in with the Bishop." I knew he was flying in with Bishop Fernandez. Then I gave him the book and I asked him to give us his blessings for all my family and my life. I also wanted a blessing for the work I was now starting. He gave his blessing. And he gave me a wink after the blessing. The Pope also said the Church was going through a very difficult time and we needed to pray a lot and not to be afraid. He then left and I opened my eyes to find I was back in my bedroom with my friends.

They were crying and praying, scared, very afraid. Shocked. They asked what had happened. My friend said, "One moment your body was elevated, your body was on top of the bed, but elevated, and there was a space between your body and the bed. And when you returned your body dropped back fully onto the bed."

Katya says she then asked her friends not to tell anyone until she had a chance to speak with Father Renzo. They all then went to sleep. Father Renzo arrived with Archbishop Fernandez the next day.

Mike and I both, at different times, interviewed Katya's friends about the incident and asked them about the bed during the time that Katya was undergoing this experience. They both agree that the indentation around Katya's body disappeared. She was lying on it, but it was as though she had no weight. Her body did not push down into the mattress.

After that, there was a second occasion of a bilocation with Pope John Paul II. Over the years, Katya had told Mike and me on numerous occasions of the problems she was having with a particular archbishop. She felt persecuted because he wouldn't allow her to publish the messages she was getting from Jesus.

Katya: "One night when we were in Merida I was sleeping in my room and I felt the presence of someone. I turned around and found myself in front of a tall window with light coming in; in that light, I saw the figure of a person in front of me. He said, "Be calm, it is me. I want to tell you that I sent a person to talk about you with Archbishop Berlie. Don't be afraid. I will always look after you." My visitor, who just appeared in my room, was Pope John Paul II. Days later, David Lago, a very close friend of ours, had a meeting with Archbishop Berlie and started talking about me and told Archbishop Berlie that the Pope supports me. Berlie asked, "Why do you say that?" David responded, "Because the Pope is looking after her." Berlie questioned why he would say such a thing. "Has not a person from Rome, from the Pope, come to talk to you about Katya?" David asked. Archbishop Berlie confirmed that a person from the Vatican had come to see him and had mentioned Katya in passing.

To Mike and me this is very important because the question of bi-location is one which is difficult to support without evidence even though we know it is accepted within the Church under certain circumstances. We do know that one of the Pope John Paul II's right-hand men, Archbishop Francesco Monterisi, a Secretary of the College of Cardinals and Secretary of the Congregation of Bishops of the Roman Curio, did visit Archbishop Berlie and that permission was then given for Katya to spread the messages. Her story continued:

Katya: "The Bishop had sent the books to Rome so they could be studied by a university. They studied my books and Monsignor Berlie said, 'If the books have a contradiction or an error, I'm going to prohibit them.' The university sent a document to Monsignor Berlie saying they had studied the books and they had found no errors."

Katya's happiness with her ability to publish and distribute the material she had received from Jesus was however, short-lived. When Pope John Paul II died in 2005, Archbishop Berlie promptly re-imposed his prohibition of Katya's work. It was clear that the protection that John Paul II promised Katya was indeed there in his lifetime but was not respected by Berlie upon his death.

After all the talking and filming was done Mike put his arms around Katya and gave her a big hug. He thanked her for everything. I watched the parting gesture knowing that this would be the last time they would ever see each other again. The end of a productive relationship from which so much good had come was now in sight. I was profoundly moved. I had introduced them to each other 20 years ago and they had become very close friends. Katya had allowed Mike to present her amazing story to a world of twenty nine million viewers. The Fox TV Special, 'Signs from God' was the first time ever a person had been filmed before, during and after receiving the stigmata. Mike's name will forever be associated with Katya and with the incredible international response to the broadcast.

CHAPTER 41

REBUILD MY CHURCH

I was traveling by car through an idyllic mountainous region about 150 kilometres north east of Rome. It took longer than usual because many roads were blocked by rocks and rubble, fallen remnants of collapsed buildings. Avoiding the debris demanded many detours. Just outside the town of Amatrice I was stopped at a police road block. I wondered what was going on? I was told that what I was seeing was the aftermath of a series of powerful earthquakes which had devastated the whole area. Two historic towns, Amatrice and Norcia, only about 10 kilometres apart, were badly affected by the 6.6 magnitude earthquake which struck on 30th October 2016.

Even though three years had passed I could see devastation everywhere. I saw a very prominent Church in ruin but for the main façade which was being held in position by elaborate buttressing and scaffolding. It was boarded up and guarded by police to protect it from scavengers and vandalism. A surly local shop keeper told me that the town had 10 historic churches before the earthquake. I asked him if could he show me on a map where they were so I could find them.

"What is the point. They don't exist anymore. They are just a pile of rubble."

The neighbouring town of Norcia was also devastated. This very old town goes back to Roman times and is surrounded on 3 sides by the original Roman wall. Inside the wall was a beautiful typically medieval town. It was virtually deserted. The only sounds I could hear were coming from the town's Cathedral in the main square which lay in ruins. It was not the sound of bells but of pneumatic hammers the workmen were using to break up the fallen stone walls and the engine of a crane lifting rubble onto dump trucks.

There were a couple of wine bars and coffee shops open to serve construction workers and the odd tourist, now a rarity, keen to see what had happened to the famous little town. Norcia is historically connected with its most famous resident, St Benedict, the patron saint of Europe, who was born there in the 5th century and is still honoured in the main square and in the Basilica, built on top of where he once lived and named for him.

Most of the houses and shops seemed to have suffered little damage in the earthquake. But strangely, every one of the town's historic churches were either severely damaged or totally destroyed. The greatest loss was the almost complete destruction of the Basilica of St Benedict in the main square. For centuries the Basilica had been home to an order of Benedictine monks who offered daily Mass and served the community with all manner of charitable works.

On the 21st September 2019 I met with the prior of the Benedictine Monks of Norcia, Fr Prior Benedict Nivakoff, an American priest. We sat in a little coffee shop off the main square and he relayed something of the terror that he and the people of the town had suffered during the earthquake. He and his fellow Benedictine Monks were dismayed at the almost complete destruction of their beloved Basilica that had stood there since before the middle ages. They were given a short time to remove their personal belongings before the army and state authorities moved in to secure the premises and to deny anyone access to the ruins.

Fr Prior Benedict told me that the town had 12 historic churches and all of them were either damaged or destroyed. He said it seemed like the churches had been singled out for demolition. It was uncanny. The historic civic building next to the Basilica escaped damage almost completely. At the time of the earthquake the Basilica and the cathedral were the only churches that were being used as churches. The rest were either used as museums or art galleries. Or just boarded up. He said that only about six people used to come to daily Mass in the Basilica. The Cathedral fared not much better with about twenty on a good day.

"There is no faith here anymore. Like this whole country. This once very Catholic country is Catholic no more. It has become very secular. There are no new vocations to the priesthood. What is even worse is that people don't seem to care about how bad the Church has become. People don't care about the state of the priesthood, the watering down of Church teaching and the general loss of faith. Homosexuality, the infidelity of many priests to their vocation and priests having women have become accepted in the prevailing culture."

Italy has suffered a series of devastating deadly earthquakes in recent years in which many people have been killed and churches in particular, often exceptionally, have been destroyed. Some estimates put the number of churches affected as high as 700.

- In 1997-98 the Marche and Umbria earthquake severely damaged St. Giovanni Battista at Apagni and Santa Croce at Case Basse.

- In 2002, an earthquake measuring 5.9 hit the Molise and Apulia threatening the collapse of San Giuliano Di Puglia Campobasso.

- *In 2009, a 6.3 magnitude earthquake hit the Abruzzo region. 306 people died. The picturesque medieval buildings of L'Aquila were severely damaged and so too were some of the Catholic church's architectural treasures; the largest renaissance church, San Bernadino, the church of Anime Sante, the baroque church of St Augustine, and the 13th century Basilica of Santa Maria di Collemaggio were all struck. More than 700 churches in the region suffered damage, some beyond repair.*

- *In 2012 two major earthquakes struck causing 27 deaths, widespread damage, and the destruction of all the churches in the provinces of Modena and Ferrara, including the Cathedral in Mirandola.*

The Benedictine Friar discussed what has become clear to all. Historically the civic life and development of all the little towns and villages devastated by these earthquakes had originally orbited around their churches. Their churches expressed their particular histories of art, of faith, of the generations of people baptised and buried by them. But as the people refused the faith of their fathers the Church has essentially become obsolete. In this new post-Christian world it has collapsed long before the first tremor struck or the first wall cracked.

He said that he was not constrained from speaking about the crisis of the Church in Italy because his Benedictine order was outside of the Italian Church's direct control.

"I don't think it is any coincidence that all these churches have been destroyed. I think God is telling us something. He wants his Church to be rebuilt not in physical terms but in spiritual terms. As for the physical rebuilding, the churches in effect belong to the State here. I don't think they will ever allocate the funds that would be needed to do the rebuilding and restorations. This may take hundreds of years. God is telling us something: Rebuild my Church."

CHAPTER 42

HAVEN'T I HEARD THAT SOMEWHERE BEFORE?

Of all the Catholic saints, the most widely known outside of the Church, by protestants, schismatics, pagans, agnostics and atheists alike, is St Francis of Assisi. In part this popularity is attributable to there being so much written about his short life in every century since his death in 1226. Blessed Thomas of Celano was given the task, by the reigning Pope, of collating the testimonies of St Francis's friends and religious brothers and shortly after the saint's death his first biography was completed. In the early stages of Francis's conversion he found himself returning from Rome and passing by the rubble and remains of what was once a Church. Blessed Thomas described the scene:

'Changed now perfectly in heart and soon to be changed in body too, he was walking one day near the church of St. Damian, which had nearly fallen to ruin and was abandoned by everyone. Led by the Spirit, he went in and fell down before the crucifix in devout and humble supplication; and smitten by unusual visitations, he found himself other than he had been when he entered. While he was thus affected, something unheard of before happened to him: the painted image of Christ crucified moved its lips and spoke. Calling him by name it said: *Francis, go, repair my house, which, as you see, is falling completely to ruin.* Christ had spoken to him from the wood of the cross in a new and unheard of miracle? From that hour on, his soul was melted when his Beloved spoke to him…. Indeed, he never forgot to be concerned about that holy image, and he never passed over its command with negligence. Right away he gave a certain priest some money that he might buy a lamp and oil, lest the sacred image should be deprived of the due honour of a light even for a moment. Then he diligently hastened to do the rest and devoted his untiring efforts toward repairing that church. For, though the divine command concerned itself with the church that Christ had purchased with his own blood, Francis would not suddenly become perfect, but he was to pass gradually from the flesh to the spirit.'

Thomas of Celano, *Second Life of Saint Francis*

Francis interpreted the command which he heard clearly with his own ears, and which was repeated three times, by assuming that Jesus wanted the crumbling church building to be re-built. To do this he took fabric from his silk merchant father's shop and sold it to buy building materials. His father reacted predictably and accused his son of theft, cowardice, profligacy and insanity and dragged him in front of the town's bishop to renounce him as his heir.

The bishop gently told Francis to return the money and assured him God would provide. Francis gave back all the money, and all the clothes he was wearing and, standing naked in the public square, proclaimed, "Pietro Bernardone is no longer my father. From now on I can say with complete freedom, 'Our Father who art in heaven.'"

Francis continued his work of obeying the mystical request. He begged for stones and rebuilt the San Damiano church by hand. It took a few years, and few repaired churches, before he made the realization that it was the Church with a capital 'C', Christ's mystical body, a church plagued by sexual scandal, heresies and confusion, which truly needed repairing. Francis never preached reform or revolution. He implored a wholehearted return to God and complete obedience to the traditional teaching of the Church. He insisted on complete reverence for, and faith in, all the defined dogmas of the Catholic Church and especially the real presence of Jesus in the Holy Eucharist.

The Catholic Church was in need of great reform in the 13th century. Confusion about the Church's teaching was widespread, and many clergy members were not living a life according to the Gospel. In response to the crisis, Pope Innocent III convoked the Fourth Lateran Council to address the issues and institute reforms. Bringing them into practice was a formidable task and it is providential indeed that at that exact time a saintly man made a journey to Rome to meet the Pope. Francis was seeking canonical approval for his newly established Order of Friars Minor, on what we today call the Franciscan Order. Pope Innocent was highly sceptical of Francis, a scruffy, scrawny pauper wearing a sack with sandals on his feet. He dismissed Francis along with his plan for his community known as the Rule. They likely would bring him nothing but headaches much like the heretical eccentrics known as the Albigensians. Then, according to a popular legend recounted by Leon Le Monnier in his History of St. Francis of Assisi, Pope Innocent had a powerful prophetic dream. He had gone to sleep highly anxious and concerned about the misfortunes of the Church when he dreamed that he saw a beggar holding up the Lateran Basilica which, like a church struck by an earthquake, was tottering and about to collapse. Then he saw himself paternally embracing the mendicant called to this high mission.

In the following days he completely unexpectedly yet unreservedly approved the Rule of Francis saying, "This is indeed the man who has been called to sustain and repair the Church of God."

Francis' order of mendicant friars was approved in 1209 and, by their missionary zeal for the traditional teaching of the Church had a great impact on the world helping to reform and save the Church at a crucial time in history.

On 4 November 1997 Katya Rivas received, and wrote down as it was being relayed to her, an unexpected message from an unexpected person: Saint Francis of Assisi. She published it in a collection of messages called Crusade of Salvation. (Message no. 128). This is what St Francis said:

"Humility is the conquest of the soul, and it is a light from God, and not only a light, but also a conquest. He, who does not know what it is to truly humble himself, persists in many errors and blindly holds in great esteem, that which is actually vile and contemptible.

Consider that to humble oneself before the Cardinals of the Church and the very Pontiff as I did, is no big matter for the one who regards himself as a sheep of the great fold. But that act of humility remains before God throughout the centuries and through eternity, so as to witness to the truth possessed, believed and esteemed. And the truth in these cases is to hold oneself as good-for-nothing who obeys God and goes to the encounter of the great ones of the Church with a simple attitude, as mine would say: with a Franciscan attitude. Therefore, self-knowledge produces humility, while self-esteem, which is pride, produces confusion and ruins what little truth may already be found in the soul.

I am Francis, the light of Assisi, the torch of Christ, the star on the firmament of the sublime Mother Church. Go with confidence, sister Catalina."

Two of the 10 historic churches in Amatrice Italy, destroyed by an earthquake on 30 October 2016.

Two of the 12 historic churches in Norcia Italy, damaged or destroyed by the earthquake on 30 October 2016.

The main Cathedral in the town of Norcia being restored after earthquake damage.

A monk who was present when an earthquake struck the Basilica of St Francis in Assisi on 26 September 1997 gave his account of what happened.

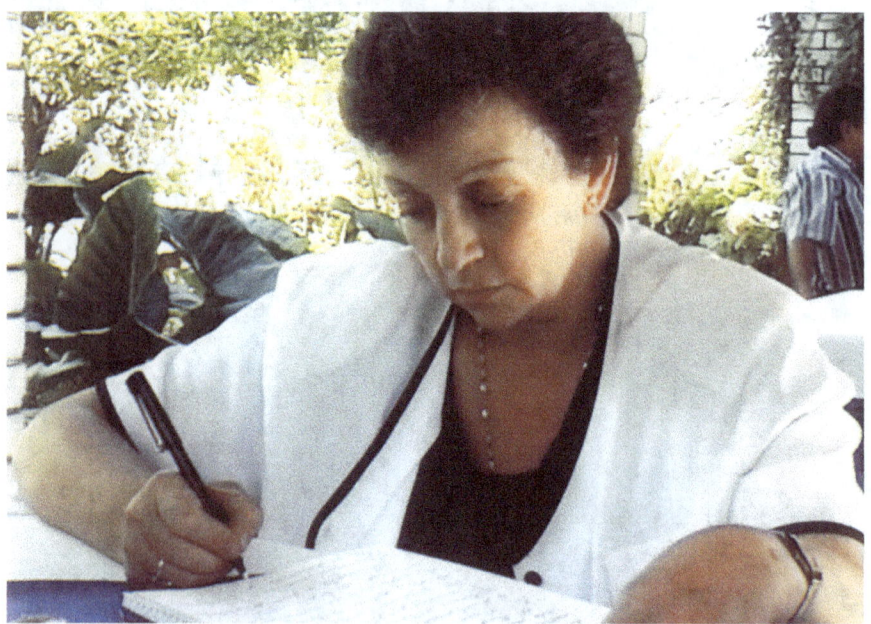

On 4 October 1997, Jesus dictated a message to Katya addressing the earthquake destruction of the Basilica in Assisi. "Did I not say to St Francis, 'Rebuild my Church!'...This is the start of many other religious monuments which will fall down".

CHAPTER 43

EARTHQUAKES

The Benedictine friar I had spoken to in Norcia analysed the catastrophic collapse of churches throughout Italy in a way unheard of in media. His view echoed those famous words heard in a mystical encounter by the most famous of all medieval Italian youths: "Rebuild my house. Rebuild my Church." What the friar didn't know is that separated by centuries and continents, another mystic, not medieval but contemporary, not a young man but an old woman, received a message about earthquakes in Italy from Jesus, and I was there to see it, hear it and film it.

Professor Angelo Fiori from the Department of Legal Medicine of the Gemelli Hospital in Rome had not been able to get a genetic profile from his testing of the first sample of material I had taken from the bleeding statue in Bolivia. This baffled him. He asked me if I could provide another sample, perhaps fresher, if and when the statue 'bled' again, to conduct further DNA testing.

With this intention I travelled to Cochabamba in October 1997. When I arrived in Bolivia, I asked Katya if she would ask Jesus if he would help me out and make it happen so that the statue would bleed.

The next day, 4 October Jesus responded, but not as I intended. While waiting for our lunch order to arrive I noticed Katya take out her notebook and begin to write. I started filming. After writing for some time uninterrupted she began drawing a crude line drawing consisting of a series of connected dots. When she finished she read it out, and others at the table translated from Spanish.

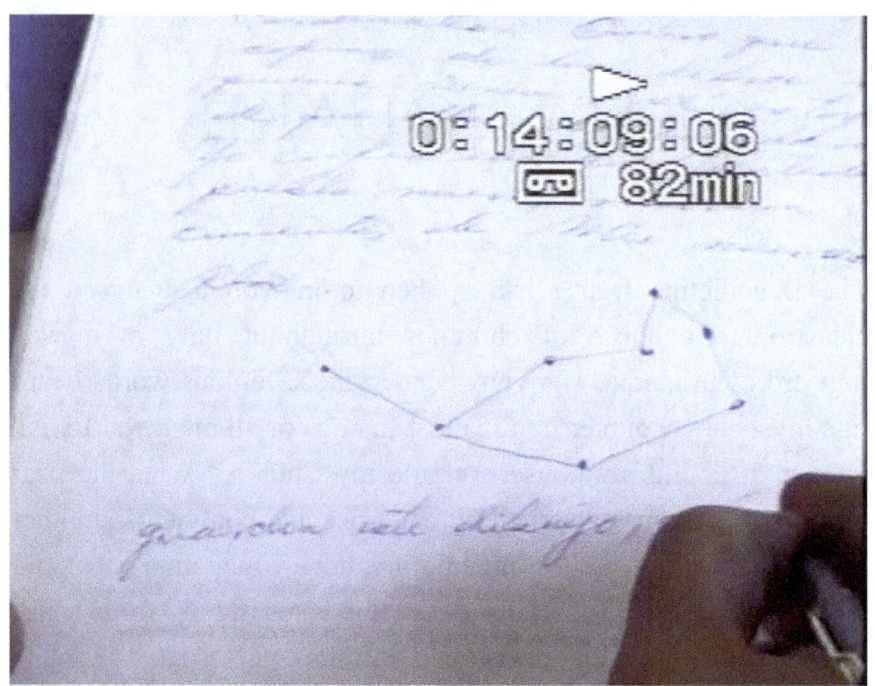

"My dear child. I want to talk to my children about the samples. It is not necessary to obtain a fresh sample because there is something that brings all my things together. The Basilica of St Francis has begun to fall down. But why? Let us see the places that have been destroyed. The tympani. What does it mean?"

At this point Katya interrupted. "What is a 'Tympani?" she asked. She looked enquiringly at the people around the table. Clearly she had no idea what the word meant. She shrugged and carried on reading the message.

Later the entrance was destroyed and then the interior. Did I not say to St Francis, "Rebuild my Church?" What does man want? Does the head of the Church want more proof of the truth of what is coming to you? The church that St Francis began to rebuild was a metaphor for today. But today the falling down, the destruction of this basilica is another metaphor. Please I need you. Please help me. Tell my children that this is the beginning of other religious monuments that will fall."

The message concluded with an instruction regarding the cryptic drawing.

"Keep safe this drawing. So will be the chain of what will fall down".

We all listened attentively and examined the drawing. Although it was a series of dots joined by lines there was no indication which dot came first, which direction the 'chain' took. The image had presented itself on the mind of the mystic in a single completed frame. She assumed it to be some kind of map, indicating the places where there will be destruction by earthquakes in the future.

Then discussion arose about the meaning of the word 'tympani'. Jesus had mentioned it and then in the very next sentence specifically asked, *"What does it mean?"* One person said it is part of the ear. Another said it's an architectural term. It turns out that both were correct.

The tympanic membrane is a thin, highly sensitive skin in the inner ear which vibrates when sound waves strike it. The sounds are then passed to the brain for interpretation and meaning. It is otherwise known by the more descriptive common term, the eardrum. 'Tympani' is the Greek word for drum. Damage to it results in loss of hearing, even complete deafness. If someone speaks you cannot hear. Perforation also often results in infection.

Architects know the tympanum as the decorative feature above the lintel of a door or an entrance typical in gothic churches. Often it is semi-circular and rests of the posts on either side of the door. Sometimes it is triangular. Invariably it depicts religious imagery as a bas relief sculpture designed as a visual lesson in one or other aspect of the faith.

In the message to Katya it is the impact of the earthquake on the 'tympani' first that is dictated by Jesus for a reason. He goes on to say *"Did I not say to St Francis, Rebuild my Church?"* The structure of that sentence clearly implies that there has been a failure to hear what he said and asked of St Francis. His call has not been heard.

Once again, almost a millennium later, he is trying to reach the ear of the Church so that it will hear his call again for the rebuilding of the Church.

In trying to make sense of what had just happened I reflected that Jesus said that there was a connection between the outcome of the DNA testing I was doing on his blood and his plea to the Pope to hear his emphatic call to rebuild his church which was in ruin. It would be, he said, through the shaking and destruction of Churches and religious monuments by a series of earthquakes that the Pope and the Catholic Church would recall the instruction he gave to St Francis in the 12th century to rebuild his Church which was then in ruin.

In the message Jesus gave to Katya it is clear that he wanted a rebuilding of his Church founded on his teachings. Jesus was calling for a renewal of the Church and was seeking help and cooperation in what he had planned. Because the message began by reference to the blood sample, and that I was fortuitously there to record it on film as it was being delivered, I am convinced that the results of testing of the sample is somehow going to be part of Jesus' plan to rebuild his Church.

Unbeknownst to any of us at the table, certainly not to Katya, and not surprising considering that international news was slow to reach a small town in central Bolivia prior to the internet, there had been a frightening geological event a few days before in Italy. On the morning of 26 September 1997 two earthquakes struck Assisi in quick succession and there was widespread devastation. Four people, two of whom were Franciscan friars, were killed in the famous basilica of St Francis in Assisi. The Basilica, parts of which dated back to the twelfth century, had been almost destroyed. The tympanum was the first to collapse. The vaults bearing the Giotto frescos and a series of biblical scenes attributed to the Roman artist Pietro Cavallini lay in the rubble.

I recalled the last lines of the message ending with an ominous prediction. The destruction of Basilica of St Francis in Assisi was just the beginning.

Two weeks after that lunch with Katya I went to Assisi in Italy to see the destruction for myself and to interview any witnesses. One, a Franciscan friar was sleeping in the Friary when the tremors shook the entire Basilica complex.

"At 2.30 in the morning on 26th September I was awakened by the sound of very loud bang. Everything around me began to shake. The other friars and I ran out of the building. We realised it was an earthquake. Around 23 minutes to 11 in the morning some 20 technicians had gone in to see what damage was done to the frescos when the second big earthquake came. Again there was another big bang and a tremor shaking everything. Things began falling from the ceiling. A panel above the altar of the Basilica fell down with some paintings by Cimabue. Very, very special paintings. Another panel fell down right over the entrance to the basilica. It was a great shock. We have had other earth quakes in the past since the basilica was built over 700 years ago. There never has been anything happen to the basilica. So we thought that nothing could happen to it. We could not believe it. We were not sure whether it was a dream or reality."

CHAPTER 44

WHAT DOES HE WANT?

'Everything begins, one might say, from the heart of Christ who, at the Last Supper, on the eve of his passion, thanked and praised God and by so doing, with the power of his love, transformed the meaning of death which he was on his way to encounter. The fact that the Sacrament of the Altar acquired the name 'Eucharist' – 'thanksgiving' – expresses precisely this: that changing the substance of the bread and wine into the Body and Blood of Christ is the fruit of the gift that Christ made of himself, the gift of a love stronger than death, divine love which raised him from the dead. This is why the Eucharist is the food of Eternal life, the Bread of Life. From Christ's heart, from his 'Eucharistic Prayer' on the eve of his passion flows that dynamism which transforms reality in its cosmic, human and historical dimensions.'

Pope Benedict XVI at Basilica of Saint John Lateran, 23 June 2011

One of the many books of compiled messages received by the mystic Katya Rivas is called 'Door to Heaven'. In it we read the teaching Jesus gave on 8 August 1996:

'I want you to know that a long time ago, the armies of the West came seven times to My tomb looking for Me. Their efforts were not rewarded, but this did not stop their intensive search for Me. My light opened a path through the storm clouds raised by the heresies that offended Me in the sixteenth century. Later the revolutionary guillotine that was beheading My priests, heralds of the faith and ministers of the Sacraments, was powerless to

put an end to the prestige and sovereignty of the one they called infamous: Me.

But, where do I live?

In the Blessed Sacrament.

That is how I have been instructing My Church and, therefore, it is not strange that all Catholic worship revolves around this Sacrament of Love, to glorify My Father and His Son, the God-Man.

Teach, My children; speak to the people about My presence at the Altars. Tell them about My infinite love and tell them how much they are missing when they do not receive their God as true nourishment...

For men of faith, My Eucharistic form is a gateway to Bethlehem where they adore Me, like the angels, the shepherds and the Magi. For those enlightened with the brilliance of the faith, in each altar where sacrifice is elevated, a new Calvary is raised and with feelings of profound adoration, those words are repeated:

"Truly, here is the Son of God."

The abandonment of the visits to My Sacrament of Love, the scarcity of communions, the excessive regard for other people's opinions by which many are embarrassed to approach the table of the angels, arises from a lack of faith. Because if they would realize, if they were told with conviction that the infinite Majesty of their God, with His sublime courtiers, resides in their temples, men would be anxious, desirous and solicitous to merit the honour and happiness of speaking, receiving and living with their Lord.

Where can we find the remedy for this faltering or lukewarm faith? In the same Form, because it sustains and increases faith. It is virtue that infuses supernatural faith. Grace that is the root and foundation of justification, grace that comes with the Sacraments and you increase with the reception of them. However, if the faithful observe that the priest is a routine celebrant, if they do not notice true fervor and humility at the moment of the Consecration, they will find it hard to believe that this will increase their faith.

Speak about the blind man of Jericho, who not only received the ability to see but had Me for the first object of his vision. And in the healing of another blind man, note that the strength of his vision gradually increased as his faith increased. In a similar manner, as men receive Communion, they get to know Me more intimately and listen from the depths of their hearts to My supreme teachings. I, being the Teacher, increase the light in the minds of those who receive Me, so that they can penetrate more into the depths of My doctrine. This doctrine, a seed from Heaven, can fall, as I have indicated, sometimes on fertile soil and will produce a hundredfold, while at other times on the road and other times among brambles and thorns... Therefore, this growth of faith, sometimes subjective and other times objective, depends largely on their cooperation with grace.

Happy are My Apostles, who after their first communion at the Cenacle said: Now we understand everything You tell us... Happy, Thomas, the doubter, who touching My wounds burst forth into an intense act of faith and love. Happy are the disciples of Emmaus, who recognized the pilgrim as the One who lit up their hearts... So happy are the poor and ignorant, according to the world, who through continual contact with Me have learned celestial wisdom, that which communicates faith, enlivened with the reception of the sacred banquet. These privileged beings not only have faith, but live justly, satisfied in practice with My prescriptions. Blessed are they! More blessed still are those whose faith in My Sacrament has been rewarded with miracles.

CHAPTER 45
GABRIELLE

My wife Gabrielle would often come into my office to see what I was working on. I showed her an email I had just received from Dr Robert Lawrence, the Californian Forensic Pathologist who had worked on my cases. He had just finished reading *Unseen- New Evidence, The Origin of Life under the Microscope*. This book that I wrote with Lee Han presented the argument that the Darwinian theory of the Evolution of Man was no longer tenable in the light of the scientific findings of the Eucharistic host-to-heart cases, cases which Dr Lawrence help build with forensic analysis.

Dr Lawrence wrote:

"I am not religious, as you know, but find the book to be a convincing argument in favour of a Creative Agency over stepwise evolution. The book very nicely points out the inadequacy of science and Darwin's Theory in explaining how or why man was created. It makes a strong argument for Divine Creation, even without the host-to-heart cases which, when added to the equation, make Darwinian Evolution an inadequate explanation of how the world and its inhabitants came to be."

This scientist acknowledged that my argument knocked Darwin's theory of the evolution of man off the pedestal it had occupied for 150 years proclaiming that God was unnecessary to explain our existence.

I showed Gabrielle the interview from the documentary I was editing of Dr Lawrence saying he was the son of the world famous Dr Earnest Lawrence, who won a Nobel Prize for inventing the Cyclotron which split the atom and started the nuclear age that we now live in. He had wished his father was with him on that day to give us some good advice about his recent discovery that human tissue had spontaneously and mysteriously arisen from nothing.

Gabrielle watched intently as Dr Lawrence said, "It is not within our realm of understanding. I would have no explanation for it. I have no axe to grind. I think it would be wonderful if this turned out that this was real. It would be spectacular. It would set all of us on our hind heels. It would make us rethink all our concepts, I am talking about scientists. My father was a scientist. I think it would be wonderful for finally some supernatural or miracle type event could ever be actually proven. I think it would be of great benefit to mankind."

She then turned to me and said, "Ron, what a great ending! It will be of great benefit to mankind. And this will be as a result of your work. Well done."

Gabrielle then walked out of my study. It would be for the last time in her life. It was just after New Year and she had a busy couple of weeks with a full house for Christmas and New Year celebrations. She wanted to escape the family and the summer heat and so we drove to a beautiful beach and waded out together. She was tired, she said, which was unusual for her. She never complained and even her recent heart problems did not affect her tireless support of me and my work, and our family.

As we were coming out of the water back to the beach Gabrielle said to me that she was feeling really weak and she began coughing. I helped her to a park bench next to the road. Then she collapsed completely. I immediately called for an ambulance. For the longest thirty minutes of my life I held her in my arms as we waited. Eventually the paramedics arrived and she was taken to hospital. She had a severe heart attack and due to the delay had suffered brain damage. The doctors said that her brain had stopped functioning and they placed her in an induced coma on life support. She was not likely to survive. She was 64 years old.

Nothing could have prepared me for my overwhelming sense of loss and helplessness. My wife, my inspiration, and my best friend for 42 years was slipping away.

Our family came together to discuss the medical recommendation of removing her life support but the decision was ultimately mine. I had quite a dilemma. For years I had been working on cases documenting God's intervention in our world and I had the faith that God could do anything, even that which seemed impossible to science. I pleaded with God to bring her out of the coma and allow her to recover and live. I did not want to jeopardize any chance of God intervening by removing the life support apparatus. I procrastinated making a decision.

I sent an email to Katya Rivas, to ask her what I should do, and to pray that I make the right decision. She responded almost immediately:

'I am sorry to hear this news and I feel very much for you in the decision you have to take. It is very hard my dear friend, but if God has decided that she should die in this way, it is because she is ready for her encounter with him and because she has accomplished her mission in this life. We are praying for you and your family. Tell Gabrielle that I love her and that she has been a very special person for me. I am so sorry that I will not be able to see her again. Have strength Ron. These are the moments in which we must demonstrate our faith and confidence in the will of the Lord.'

I braced myself for what had to be done. As the family gathered around Gabrielle in her hospital bed for the last moments of her life I could see the tears and the grief in the faces of our children confronted with the imminent death of their mother. Wrapped around her arm was her scapular and the holy medals that she had worn most of her life and was wearing during that last swim. I held onto that warm arm for the last time. 17 January 2014 was the saddest day of our lives. But not hers.

Gabrielle had anticipated and prepared for this day all her life. We had discussed it many times. She had complete faith in the Christian teaching that death is not the end of your life, and that on this very day she would meet her Creator face to face and be invited to take her place in an eternal kingdom, a paradise without end. Gabrielle had confidence that she would go to Heaven after she died because she had spent her entire life seeking to know God and to carry out his will.

After her own sister died she had written a consoling letter to her niece about being a mother and getting to Heaven.

'I have been reminded of some things that helped me when I was in the thick of raising children, juggling relationships with husband, in laws, family and friends and now for the first time in my life I've sort of got time to step back and re-assess and re-group as it were, and I can definitely put myself in your shoes. The first thing that was told to me emphatically and repeatedly that started it all off for me was, *Seek ye first the kingdom of God and all else will be added unto you.* If Our Lady, the mother of God, who without doubt is co-redeemer with Our Lord of the human race, yet was hidden said to Luke writing the Gospel when he wanted to write about her, "write about my Son," then don't be afraid of what seems like a life of lack of recognition and even obscurity and just service and giving. I now embrace and rejoice in the hidden life of bringing souls to God and putting up with being misunderstood and even persecuted for it. The honours of this life are an obstacle to the life in Christ, because the latter is ETERNAL LIFE and the former is so transient and yet we invest so much time in perfecting this short existence and being noticed and recognised. It is suffering that brings us closest to God and that is why he allows it. Being a mother is a glorious way to Heaven and unrecognised, unlauded on earth and yet look at the outcome and the terrible state of humanity with mothers out there competing with the males and subjugating their nurturing instincts. Even if your life seems so much harder than others at times rejoice and as far as the others are concerned respect them where they are in their journey and wish them well. Some are more, some are less advanced in accomplishments than you, some are better at some things. So be it as long as you are doing your best for God.'

Gabrielle's life was one of 'serving and giving'. She adopted a practice of helping one person in need each day. If everyone did the same, she reasoned, the world would be a better place.

Every day, every single day, no matter where she was in the world, she went to Mass. She would kneel in church, sometimes alone, every year on a cold winter night for twelve hours continuously before the Blessed Sacrament to honour the Sacred Heart. She held a weekly prayer group for fifteen years. In addition to a life devoted to all of us Gabrielle volunteered as a Catechist teaching her faith to children in public schools for over twenty five years. She would visit the sick and elderly, particularly those who had no relatives nearby, even 'adopting' certain elderly people, doing their errands and shopping, celebrating their birthdays with them.

Spending hours comforting people in nursing homes abandoned by their own families was never time wasted for her. She would go to the funerals of people she hardly knew just to show support for a relative. On one occasion she even travelled to another city to attend the funeral of our son's schoolmate's parent, a person she had never met.

Gabrielle measured accomplishment in life by how much she could do for others and never in financial terms. In fact she was not at all interested in money and was oblivious to her bank balance. She would deliberately buy groceries which were at the end of their 'UseBy' date out of kindness to the shopkeeper. She explained that no-one else would buy those products and his business would otherwise suffer.

When I first met Gabrielle, I didn't think, "Ah ha! God has answered my prayer," because, truth be told, I had forgotten all about having prayed that prayer. In fact I only remembered it 30 years later, to the day, when Katya asked Jesus what she should give me for a birthday present. Jesus told that her I needed shaving cream, which was true but which no-one knew anything about. Then it struck me that God actually gave birthday presents when it had to do with furthering his interests. I had asked for his choice of a girl I should marry. His choice turned out to be something I needed a whole lot more than shaving cream.

From then I began to see Gabrielle in a different light. I started to see virtues and value in her to which my commercially orientated life was blind. I awakened to recognize her beautiful childlike heart. She was the most charitable and exemplary person I ever knew. I came to realise that I had married a woman who was completely unique in that she lived to know, love and serve God. She led me to the realisation of what truly mattered in life.

I had lost, suddenly, the person that I had loved so much for over 40 years of my life. I did not get the chance to say my last goodbye. She is not there when I wake up each morning. And she will never return home. It is emotionally hard to accept the loss of her as an occasion of great joy, of her finally going home to heaven.

Monsignor Micek offered a requiem Mass for Gabrielle. He spoke out loud the sentiments in the grieving hearts of more than 600 people who came to pay their last respects to a woman so loved by all.

"She became like an angel for God giving glory to him through her everyday service to the family, friends and those in need. She was fascinated with God during her life, and now she can find God's fulfilment in his presence. It was truly a grace and privilege to be in the circle of her friends."

Gabrielle quietly praying from her prayer book outside Westminster Abbey, guarding the camera gear, while I was filming the façade of the Abbey.

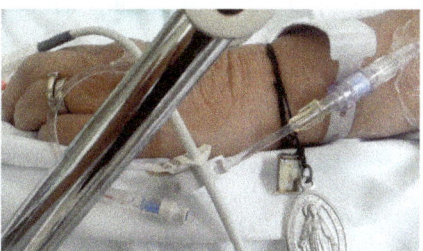

Gabrielle says her last goodbye to Katya on 7 September 2009.

In 2014 Gabrielle suffered a heart attack and was placed on life support and soon after passed away.

CHAPTER 46

HEAVEN

In 2001 Gabrielle and I accompanied Katya to Fatima in Portugal, to where the Virgin Mary appeared to three shepherd children in 1917. On the way back Katya had a mystical experience of the Virgin Mary in which Mary asked specifically that Gabrielle establish a prayer group in our home in Australia. And of course Gabrielle dutifully obeyed. Every Friday night she presided over a group she recruited, organised, and catered to personally for nearly fifteen years. It still meets.

The meetings included traditional prayers, prayer intentions and spiritual readings. At one meeting we read a message dictated to Katya by Jesus in 1996. It was a memorable evening because of the beautiful reading and the animated conversation it ignited. Everyone contributed to the discussion on Heaven, on what it must be like, on what we hope for and imagine, on what the Bible described, on what the saints have said, on what Jesus promised.

"I want us to talk about Heaven. The place you must talk about to encourage My children to work toward its conquest. Heaven, My children, is a gift so great that I desired to die on the Cross in order to open for you, entrance into it. In Heaven there is no death, no fear of dying; no pain nor illness, no poverty, no heat. There is just a serene eternal day , a continuously blooming and delightful spring... Everything you can desire is there, ...Seeing that magnificent city that is so beautiful will satisfy your sight. You will see that the beauty of the inhabitants further enhances the beauty of the city because they all dress like kings, and they are kings. The reward promised to you is not only beauty, harmony and other good things, but I, who allow Myself to be seen by the blessed ones. In Heaven, souls are certain that they love and are loved by Me. They see that I embrace them with immense love and that My love will never end. That love grows then with the conviction of how much I

loved them when I offered Myself in sacrifice for them on the wood of the Cross and I turned Myself into sustenance, into food, in the Eucharist. It is there when the soul will see that those tribulations, that poverty, those illnesses and persecutions that it considered misfortunes, were nothing more than love and the means that I used to lead it to Paradise.

Believe Me, little children, saints and martyrs say they have done little to get to Heaven. But to what does their entire suffering amount compared with that sea of eternal joys, where they will remain forever? Take heart, My children, and prepare to suffer patiently all that you must suffer in the time that is left, because everything is paltry and nothing compares to the glory of Heaven. When you are distressed by the sorrows of life, look up to Heaven and console yourselves with the hope of Paradise. There My Mother awaits you. There I await you, with the crown in hand to crown you as monarchs of that Kingdom without end."

(From Door to Heaven 30 October 1996)

Gabrielle was sitting opposite me as the words were read and she was enthusiastically full of joy at the prospect of heaven. She wished for it to come soon. She said she was ready. Reading the message now is a bittersweet experience because I will never see her sitting across the table from me again and yet I am more certain than I have ever been of anything that she is indescribably rewarded in her destiny right now. My certainty is not simply a personal opinion or morale-boosting delusional wish fulfilment. At least four people on three different occasions in three different cities have come forward to tell me of their experiences after Gabrielle died.

Our daughter Katrina was married and living in the United States when Gabrielle suffered the heart attack. She caught the first flight back to Australia to be at her mother's bedside while she was on life support. Tony, Katrina's husband, left a couple of days later. By the time he arrived Gabrielle had died. In order for him to pay his last respects Katrina requested to view the body. This is her account what happened at the funeral home:

"We were brought into a small room where mum's body lay in the coffin. We were left by ourselves for a short while. I noticed that mum was dressed in a white shroud and she had her religious scapular draped around her neck and onto her chest. But most noticeable, was a very strong beautiful fragrance coming from her body and radiating into the room. It was a sweet flower like fragrance yet of a flower I could not identify. Tony remarked that he also could smell the fragrance. "They have put some perfume on Mum," I said to Tony. "She would have hated that. She didn't wear perfume."

I stayed a while talking to mum and praying, I then lent over and kissed her forehead. Her body was so very cold. On the way out we asked the funeral director if they had put some kind of perfume on Mum. They said that they did not. They could not explain the perfume fragrance we sensed."

I recorded an interview with Mike Willesee just before he died in which he said he was consoled about what would happen after death. He used the interview to construct the following excerpt for his book, *A Sceptic's Search for Meaning:*

> 'I have now had the most indescribable experience of seeing someone in heaven. I know that is a big statement and one which will be questioned by many. But my belief in the authenticity of this experience is unshakeable.
>
> Ron's wife, Gabrielle, died on the 17 January 2014. She was a very good friend of mine. In fact, she was a very good friend of everyone she met. She was just that kind of person. She travelled with us on many of our adventures – millions of miles around the globe – while we put together *Signs From God* and our other investigations. I was a regular visitor to her home and I knew her well. She was a woman of extreme charity. She constantly went out of her way to help anyone who was in need – the poor, the hungry, the sick, the dying, the old person down the street who just needed help with their shopping. Gabrielle was just always there, an inspiration for those of us who would like to live a better life.

It has never crossed my mind that anybody would appear to me from heaven. It's not the sort of thing that happens to me. But I'm at Mass in Sydney's St Mary's Cathedral on the first Sunday of December 2016 when, suddenly, Gabrielle is there. It is an innocuous image of her that I see but, oh, so powerful: her full face, her hair and a hint of the top of her shoulders. She does not look at me and I get the impression that she is oblivious to the fact that I can see her. She appears to be listening to someone, maybe diagonally across from her, but close.

I am stunned by how beautiful she is. She was a beautiful young woman when I first met her, but she was in her sixties when she died and her looks, as you'd expect, had changed. But there is not the slightest doubt in my mind that this is my friend, Gabrielle. She appears as if she is in her late twenties and her beauty is beyond what I had encountered in life; smiling quietly with a look of contentment and relaxation.

When Ron later asks me to describe what I saw, I have trouble finding the words because the ones that come to mind, like 'beautiful', 'relaxed', and 'serene', don't seem sufficient. They are too common and overused. Then the right word comes to me in a jolt: 'heavenly'.

So there she is, almost like I am watching a wall-sized television barely two metres away, until she disappears as quickly as she had materialised, leaving me in a pleasant state of shock.

Looking at it now, I can wonder if this was a message related to my fate: Was she telling me I'll be joining her soon? Or that everything's going to be okay? That I'll be cured! But I don't ask myself those questions at the time. They seem to simply not matter. I just feel great comfort… and joy.

Six weeks later, Ron calls me with some news.

'Somebody else has just had an almost identical experience with Gabrielle as you had,' he says.

Ron explains that he has a friend from Melbourne (who I'll call 'Burt' because he's asked not to be named). Burt and his wife, Anne, were regular visitors to the New South Wales Central Coast and whenever they were around, they would join Ron and Gabrielle Tesoriero in their weekly prayer group.

When Burt rang Ron to tell him about the apparition, Ron asked Burt, who is also a lawyer, to write it down.

Burt wrote: 'On the evening of Saturday 14 January 2017 I went to bed about 11.30 pm. A few hours later, 2.30 am, while still asleep, a young and very beautiful woman appeared to me. I thought she was aged in her twenties. I could only see her from the shoulders upwards. Her face was beautiful. She was happy, peaceful and heavenly. There was a certain luminosity about her face that was hard to describe. She looked directly at me but said nothing. That vision took up the whole scene. I only knew Gabrielle when she was about 60 years of age. I did not know what she looked like in her twenties but somehow, I knew it was definitely Gabrielle. Immediately after the vision, I woke up. The whole thing was very real to me. When I woke, I could recall every detail and immediately woke my wife, Anne, and told her what had happened.'

Ron had not told Burt about my vision beforehand so there was no question of prompting. When he did tell Burt my story, Burt revealed that he had just been diagnosed with thyroid cancer and he believed that could be related to Gabrielle's appearance. That her appearance was 'a great consolation' to him.

I have never met Burt, nor had I even heard of him before this. I don't need the coincidence of his story to confirm my certainty about my vision of Gabrielle, but it is nevertheless very pleasing and uplifting news. Burt also later tells Ron that after the intercession of Gabrielle, his tumour is no longer visible to his medical team.

Two things have stayed strongly with me since that news about Burt. One is his comment about Gabrielle's appearance being of 'great consolation' to him. The words I had used in my report to Ron were that her appearance was a 'great comfort'. Different words but precisely the same sentiment. And second, he'd used 'heavenly', the very word that I had struggled so hard to find.'

I re-read the email Katya sent at the time of Gabrielle's death:

"If God has decided that she should die in this way, it is because she is ready for her encounter with him and because she has accomplished her mission in this life."

Katya's words were very different from the many condolences I had received after Gab's death. For those who believe every life has a purpose the death of anyone could be described in terms of a mission accomplished. But I believe that in the case of Gabrielle's mission the nuance might be different. In prayer I had asked God to choose the person I should marry and to give her to me as a present for my birthday. I never expected that his choice entailed a special mission of heroic holiness, patience and perseverance in preparing me for what God wanted of me.

Messages Katya has received from Jesus refer to me specifically by name in relation to the first comprehensive scientific investigation of the first contemporary Eucharistic miracle. But it has only been now that I have come to the full realisation that Jesus has a much bigger plan and that I am part of it. Something serious was and still is expected of me. I was to carry out the investigation on the Buenos Aires case and to pursue and report on the scientific findings using every skill in my legal arsenal.

I make the bold claim that the Buenos Aires case will prove to be one of the most significant events in the history of Christianity with monumental consequences for all humanity. The scientific results so far, and those one can expect from a comparative testing on the Shroud, may well lead to conclusive evidence that God is the creator of life and that he is alive in the Eucharist, just as he said he is. If the profile data from the genetic testing of the *living person* found in the communion host from Buenos Aires is the same as the genetic profile from the *dead person* whose image and blood is on the Shroud of Turin, then Jesus undeniably, scientifically rose from the dead. For a living man and a dead man to have the same genetic identity is entirely, uniquely unimaginable. Until now.

It has been an immense privilege for me to work on these cases which will provide the scientific groundwork for God to be acknowledged as the author of life and for the first chapter of the origin of human life to be re-instated. My 'working for God' was part of Gabrielle's life from the day I met her to the day she died. Gabrielle did have a "mission" in life. And she accomplished it. It is in every day since we met and in every word on every page written here and in those yet to come.

CHAPTER 47
SACRED HEART

"My love for mankind, whose symbol is the heart."
"It is MY HEART that I place in the Eucharist"

In this book I have presented cases from different parts of the world in which scientists found communion hosts have inexplicably transformed into heart tissue. In each case the heart tissue showed evidence of having suffered traumatic injury. Those words, 'communion', 'heart' and 'suffering', in association with each other, bring to mind the mystical revelations of a 17th century French saint , St Margaret Mary Alacoque. She experienced apparitions of Jesus revealing his suffering heart, saying,

"Behold this Heart which has so loved men that It spared nothing, even going so far as to exhaust and consume Itself, to prove to them Its love. And in return I receive from the greater part of men nothing but ingratitude, by the contempt, irreverence, sacrileges and coldness with which they treat Me in this Sacrament of Love. (Holy Communion)"

Jesus has not only revealed the mysteries of his Sacred Heart hundreds of years ago. He has done so again to the contemporary world by dictating messages to Katya Rivas.

"Here I am my daughter in the Eucharistic sacrament, the sacrament conceived by my love for all time. My heart which you possess in my Blessed Sacrament as a pledge of future glory and which has to be your blessing in heaven is also for you while you live on earth."

(From *Ark of the New Alliance*, 18 April 1995)

Close to the feast day that commemorates his appearance to St Margaret Mary in 1675, Jesus dictated to Katya a profound expression of the nature of his Sacred Heart and its love for mankind.

"The devotion to My Heart has a very worthy purpose, since this Heart is perfect in its own substance. The flesh of which it is formed is a seed transmitted without stain through the ages, preserved by the sanctifying action of the eternal Word, carried by the Virgin in her pure womb, which made her worthy to be conceived without sin.

The whole Trinity worked to form My Heart, manifesting in it Its grace and Its mercy. The Father created it with a perfection that

makes it worthy to be the Heart of His Son. He made it shine with the living image of His eternal beauty.

I, the Word, in becoming flesh, ennoble it, deify it; I make it sensitive to the glory of God and to the interests of men, and I make it throb in that insoluble charity that unites Me with God and men.

The Holy Spirit fills My Heart. It penetrates it and illuminates it with His divine flames that will consume it eternally. It makes of it a temple for Itself and your tabernacle; It consecrates it with Its presence; It illumines it with Its brightness; It animates and sustains it with a life without weakness nor decadence.

In this way the Father, the Word and the Holy Spirit form the Heart of the Man–God. The Father brings to this work the power that creates, the Word that healing love, the Holy Spirit that which makes fertile. Such is My Heart, such is the purest organ of holy humility which I concealed in Mary's bosom and I united it to Me, personally, when I became man.

This Heart is the generous spring from which gushed blood during My agony in the Garden of Gethsemane, when, oppressed under the weight of your guilt, I redoubled My prayers, bathing all My body in a mortal sweat. This Heart gave out drops of redemptive blood from the thorns of the holy crown and from the lashes of the whip. During three hours this heart bathed the summit of Calvary with the precious blood that paid for the rescue of the world. This Heart opened itself up after My death to pay completely with the excess of My love, a rescue that could have been paid with just one drop.

After having slept for three days in the sepulchre, this Heart awoke with all the energy and all the strength of its love, arming itself in the resurrected body of the Man-God, with the most wonderful properties that Heaven gives to My glorified body: the movement, the brightness and the grandeur of a life without end.

See in this sensitive substance, the Heart that I showed to Margaret Mary, telling her: "Look at this Heart that has loved mankind so much." This sensitive and material Heart should not be more than a secondary purpose of your devotion. There is a spiritual purpose that is the principal part: My love for mankind, whose symbol is the Heart and the flames that come out of it, and which should show you its great sentiments. This love should be for you an inexhaustible motive for admiration and gratitude"

(From *Ark of the New Alliance*, 22 June 1995)

ACKNOWLEDGMENTS

I express my sincere thanks to Lee Han for her immense contribution in helping me write this book. Lee co-wrote 'Reason to Believe' and 'Unseen' with me. Her writing and research involved understanding, debating, and then simplifying many complex scientific and religious issues that have arisen in my investigations over the last twenty years. She took these great scientific discoveries and shaped them into a great story. Her way with words, her eye for detail and design are all evident in the book and it is far better than it would have been without her. I am grateful for them. But I am even more grateful for her continuing loyal support as my friend and fellow pilgrim on the ultimate journey.

To my children, Katrina, Sophie, Alexander and Nicholas, I thank them for their support and for their forbearance in my devoting less time to them than I otherwise would have, had I not taken on the work the subject of this book over the last 20 years . I particularly thank Nicholas, a graphic designer, for his design of the book cover.

My thanks also goes to my personal assistants over many years, Lyn Petrie, Jodie Wells, Maria Dugandzich and Valerie Parkinson.

I wish to express my special thanks to William Steller, a now retired Australian ABC Television producer, who taught me many skills in filming and in documentary making and who spent a number of years working with me.

I also thank, the late Mike Willesee, Katya Rivas and those Cardinals , Bishops and Priests who have supported Mike Willesee and myself in our investigations including Cardinal George Pell , Bishop Rene Fernandez, Archbishop Edward Ozorowski, Bishop Zdigniew Kiernikowski, Mons Henryk Micek , Father John Flader, Padre Renzo Sessolo, Father William Aliprandi, Padre Alejandro Pezet, Padre Eduardo Graham , Father Mark Withoos, and Father Benedict Nivakoff.

To the many professionals that I have consulted including, Professor Angelo Fiori, the late Professor Odoardo Linoli, the late Dr. Frederick Zugibe, Dr. Robert Lawrence, the late Dr.Tom Loy, Dr. Ricardo Castanon, Dr Peter Ellis, Dr. John Walker, Dr Colin Summerhays, Dr. Richard Haskell, Dr Robert Goetz, Professor Pierluigi Lenzi, Professor Sudhir Sinha, Professor Sobaniec-Lotowska ,Professor Stanislaw Sulkowski, Dr Dadna Hartman, Dr Francesca Fontana, Dr Lesley Gray, Jorge Manuel Rodriguez and Dr.Sanchez Hermosilla, I express my appreciation.

ABOUT RON

Ron was born in Sydney, Australia. He completed Bachelor degrees in Arts and Laws at Sydney University. He was the senior partner of a law firm for many years until he became interested in examining, through the eyes of Science, claimed mystical phenomena associated with Catholic faith. His documentaries led to the making of the Fox Network television special 'Signs from God: Science Tests Faith' produced by veteran Australian journalist, Mike Willesee and broadcast throughout the Americas in 1999. He was married to Gabrielle (now deceased) and has four children.

Ron's journey of investigation over the years was first published as "Reason to Believe, A Personal Story" in 2007. New scientific findings have resulted in Ron challenging conventional theories on the origin and evolution of life. His argument, together with accompanying scientific evidence, was first outlined and sent to Pope Benedict XVI and Geneticist, Dr Francis Collins, director of the landmark Human Genome Project on 11 November 2009. He presented his argument more fully in the book, 'Unseen, New Evidence, the Origin of Life under the Microscope', published in 2013 which he co-wrote with Lee Han.

From 2013, Ron has continued work on the scientific investigation of the "host to heart" transformation in the Eucharistic phenomenon of Buenos Aires and has made a series of documentaries on the path of that investigation. The latest findings and their implications Ron has documented in this book.

www.ingramcontent.com/pod-product-compliance
Lightning Source LLC
Chambersburg PA
CBHW051420290426
44109CB00016B/1371